LIKE PLASTIC

David Flusfeder was born in 1960. He lives in London.

BY DAVID FLUSFEDER

Man Kills Woman
Like Plastic

David Flusfeder

LIKE PLASTIC

V

VINTAGE

Published by Vintage 1997

2 4 6 8 10 9 7 5 3 1

Copyright © David Flusfeder 1996

The right of David Flusfeder to be identified as the author of
this work has been asserted by him in accordance with the
Copyright, Designs and Patents Act, 1988

First published in Great Britain by
Jonathan Cape Ltd, 1996

Vintage
Random House, 20 Vauxhall Bridge Road, London SW1V 2SA

Random House Australia (Pty) Limited
20 Alfred Street, Milsons Point, Sydney,
New South Wales 2061, Australia

Random House New Zealand Limited
18 Poland Road, Glenfield,
Auckland 10, New Zealand

Random House South Africa (Pty) Limited
Endulini, 5A Jubilee Road, Parktown 2193, South Africa

Random House UK Limited Reg. No. 954009

A CIP catalogue record for this book
is available from the British Library

Papers used by Random House UK Ltd are natural, recyclable
products made from wood grown in sustainable forests. The
manufacturing processes conform to the environmental regu-
lations of the country of origin.

Phototypeset by Intype London Ltd

Printed and bound in Great Britain by
Cox & Wyman, Reading, Berkshire

ISBN 0 09 970231 2

FOR MY FATHER

ONE

THE RETURN

Let us love plastic for what it is. Let us love plastic for what it does. Let us love it in all its hard-edged, soft-moulded, brittle beauty. Let us love it for what it tells us about ourselves and our manufactured world. This shiny hard shell of a table, this soft billowing comfort of a chair, this efficient set of kitchen knives, this water glass, embossed with stars, they are plastic. This car, that television set, the hi-fi over there, the transistor radio on the window ledge, the bed, the picture, the clothes we wear, the window that the sky shines through, plastic, each of them, all of them, Carbon and Hydrogen atoms bonded purposefully together in harmonious precision, all the colours of the rainbow and the shades in between and the special tints that we have the power to make but not the poetry to name, plastic. We are not animals grubbing out our organic nests. We are beyond all that. The only limit upon us is the boundary of our imagination. We have the man-given capacity to build our visions out of a material we have made for ourselves.

Remember the days of the first plastic combs, unstable cellulose nitrate dreams fixed into the hair of Edwardian beauties; bless them, those beauties, our technology's saintly martyrs. Suddenly — a stray ember from the fire, a spark from a huntsman's gun — and the combs would combust: up in smoke the complex coiffures, scorched the scalps and faces of our darling precursors, our loves.

H. C. Willi, from his introduction to *Century of Plastic: A Celebration*, London 1962

I

HE ARRIVES BACK in town, full-haired. His suitcase contains one change of clothing, a selection of Vivien's letters (the most recent one still unopened), trophies from his travels, and the fragments of a photograph. The photograph is the old one of Danny and Sam, sent to him anonymously at his lawyer's office, the brothers in their triumph rudely ripped apart and the pieces stuck to a square of white card, pasted clumsily together by, probably, a child.

His face is tanned by sun and wind. His eyes have been deepened by laughter lines at the corners. There is a scar below his left eye. His mouth has, finally, learned how to smile. He looks strong and he feels it. He stands in the centre of the station concourse, a benevolent, crowd-jostled king travelling in disguise back to his kingdom. He would touch them all if he could, lay a regal hand on each forehead in turn, whisper a benediction, banish their suffering. The world is a good place.

Howard lingers at the avenue of telephones. A hubbub of people telling frantic lies, making final pleas, last-ditch negotiations, wrong numbers, or just — simple phrase, magnificent — calling home. He could call home, announce himself now, give them time to make their welcome party or construct their defences (he is not so high on self-exhilaration that he thinks his return will be entirely easy), that would be the considerate thing to do. He has gone beyond consideration. This is the grand thing: to return in glory and in shame to his interrupted life, to prove himself with passionate strength to be a model father, an ideal husband, a good son, a company man. The homecoming must be truer than hastily erected barricades or a freshly baked cake decorated with icing sugar in the shape of

3

his name. He wants to see the hurt in their eyes, unmaskable, the indifference, the hope. The treat of return, the full pain and joy of it.

Tube prices have gone up again. The station looks dirtier than he remembers it. He gives half his change to a beggar with a dog, the other half to a busker with an accordion. The silence on the platform offends him. Why don't they talk to one another? They can't all be frightened. He would like to teach them a song, gather them around his feet, take them through the verses line by line, try it in rounds, join in over there, you over there, yes you in the glasses and the military tie, and this time just the women, this time just the men, the boys in suits come in here, the girls in jeans take it now, this one's for those who work in offices, this one's for those who have never worked at all, this one's for the foreigners, this one's for the northerners, and everyone who's ever had a family let's belt out the chorus all together.

He is too restless to sit down in the carriage. He holds on to a rail and sways with the rocking of the train and he reads the promotional copy and the poetry and the prettily applied abuse. The stations flit by like advertisements. He tries out his smile on the passengers and most look away but a few smile back, and that means there is still hope, for them, for him, for everybody. If there are ten just men the city is always saved.

Should he get a drink? He doesn't need a drink. Cherish this feeling, dry mouth, racing heart, the sound of the blood in his ears, the station sign, doors open, onto the platform, shoulders buffeting against shoulders like runners going into the final bend of a race, the reassuring smoothness of the escalator, the shaking of his hand as he exchanges his ticket for passage out into the world. Same shapes but it looks nothing like the old world. That one was dreary, exhausted. This one hums with possibility. Things tremble, struggling to contain the energy inside them.

The walk home. Familiar unfamiliar smells. Cherry blossom on garden trees. Car fumes. Dog shit. Neighbourhood flowers. The urge to run lifts up inside him and it's a double urge, to run away, again, evade this confrontation, for ever, and the urge

4

to run home, shatter this distance between them with the power of his muscles, to the door, bang on the door, it opens, consume it all in one moment, hug them all, feel their bodies against his, the looks in their faces, the tears. He stops. He reminds himself of his poise, he is the king back from exile, he is the hawk swooping back to the nest, he is the great explorer returned from his foolish, necessary journey. He gives himself confidence by smiling at his strong tanned reflection in the wing mirror of a polished vintage car. He goes on. Hedges, trees, an elderly woman smoking a surreptitious cigarette, horse-head pillars, a pair of blonde schoolgirls skipping hand in hand but always avoiding the cracks in the pavement, a gravelly drive, a burly man sadly removing the Beware Of The Dog sign from his front lawn. A pair of closed electric gates. He is home. He is there. Here.

He had forgotten about the gates. His memory has no place for them. In his memory the path is shorter, the door is open, his family are gathered, photogenically huddled, on the welcome mat. The meeting has to be at the front door. He can't let them see him like this, electronically diminished, his voice growling out of the console in the hallway. Howard climbs the gates, he has to, suede shoes scrabbling, slipping on wrought-iron curlicues, he heaves himself up, pulls, strains, nearly falls backwards, grips the top, tugs himself over, there's the light shining behind the purple curtains of his son's room (Simon can't still be . . .? Not possible. Vivien has written of his triumphs), that inevitable moment of agony: his balls jammed at the top, his legs stupidly dangling either side — are the neighbours watching? are the police already on their way? — he pulls his left leg over, crashes the shin against an art nouveau twirl, winces, and holds; he's in control now, a year ago this would have made him sweat and lose his breath, two years ago the action would have been beyond him entirely; and now he drops lightly onto the drive, hears a crack in his knees, feels a pain in both his ankles, slowly stands up, no damage done. He walks to the door limping a little from the pain in his balls and the bruise to his shin. He knocks. He pushes his hair back away from his eyes. He waits.

He knows it will be his wife who answers the door. He had thought to prepare himself with a clutch of flowers. *Here*, he would say, *for you*, then stepping inside he would stretch, yawn, wait for the silence to be filled, for her body to find its old customary place within his. Flowers would have been laughable, adolescent. He does not think so little of her to pretend that the wounds can be sealed with flowers.

The door opens. His new smile opens along with it.

She is older and heavier than he remembers. And more beautiful too. There is little grey in her black hair. Her skin is pale, her green eyes glance at him with a fugitive look that is new to her, as if she needs to look at everything twice before she can be certain it is what it claims to be.

— Howard?

— Love.

Her face is still good. High cheekbones, high brow, rich red unpainted mouth, a fine nose that he has always liked to think of as aristocratic. There is more flesh than there used to be beneath her chin. Her body is foreign to him. It is hidden under a blue housecoat all the way down to her pale calves, her perfect feet, with nails surprisingly decorated with chipped red paint.

— When did you start painting your toenails?

— What? I don't. Oh. A game I played with Ruthie the other night. That's all. We were being silly together.

She is not answering his smile with one of hers. She is looking mistrustfully at his hair, at the suitcase in his hand and rucksack on his back, at the loose hippie traveller's clothes he wears, at the colour of his skin. This moment could last for ever and it would kill them both.

— Can I come in?

— Yes. Of course.

She steps aside to let him through, the brief reassurance of politeness. They walk through the hall, same pictures, same mirrors, same stairs, same carpet, same kitchen, same table. They sit in their accustomed chairs.

— Would you like anything? Some tea? A beer?

— A beer would be nice. Thank you.

She fills one glass with beer and another with gassy water. They sit opposite each other at the kitchen table. He had not counted on her suspicion, her reserve. Stupidly he had thought that his arrival would be enough, to send the situation spinning,

6

out of their control, and all they would have to do would be to hold on to it and to each other and wake up panting in the morning, together. It is up to him, he is the criminal.

— You look well.

— I do? Thank you. So do you.

It annoys him that she hasn't commented on his hair. It is his pride. The house feels empty. He wishes the children were here, to scream hate at him, to throw things, to forgive him everything because they have to, and then she would follow.

— You've been busy with the garden, I see.

Abruptly she gets up, pulls the blinds shut over the french doors to the garden, closes the kitchen door, sits down again. He is jolted by the sudden awareness that she is desperately nervous. He reaches for her hand, she pretends she was just about to move it away anyway and she takes too long a draught of her water which makes her hiccup.

— How, why . . . how long have you come back for?

— How long? For ever. If that's . . .

— If that's what?

— Acceptable, I suppose.

She nods. Patiently, inappropriately. As if he has just said something quite clever, but really rather obvious. He starts to try out his smile again but she's not receptive to it and he shouldn't use it too often because then it would turn into a selling tool and he would be no better than Charlie.

— How is Charlie? Flourishing I suppose.

She says he is with a nod of her head and a shrug of her shoulders, and as she restores the lapels of her housecoat to their prim position Howard suddenly feels sick. This is going wrong. It was an awful mistake to ask about villainous Charlie before asking about Simon or Ruthie or Dinah or, properly, herself.

— Look. I'm sorry. I know how hard this is for you —

— Do you? (Mildly said, almost uninterested.)

— Yes. I do. And it's hard for me too, maybe not as, but difficult all the same. All these things . . . How are Simon and Ruthie? Will they be glad to see me? How's my mother? I feel so bad about her. And how are *you*? Truly.

That new shrug again. Makes him want to shake her, it's so

7

mousy, uninvolved, self-belittling. Briefly he considers raping her.

— Look.

He reaches for the hand again and he's too quick for her this time, he's got it in his, he pulls it closer, clasps it a little too hard to show her the urgency of his feelings, strokes the finger that still wears her wedding ring. She is not crying. This is somehow both disappointing and gladdening.

— I can't, I don't know, I. Vivien. Viv. I want to come home. That's what I want.

— You want.

— I know things aren't going to be the same. I wouldn't pretend that. How are the children? Will they be glad to see me? Thank you for your letters. If it hadn't been for them I don't know.

She is frightened. He realises this with a quick sad clarity. She tries to tug her hand away but he won't let her. A solitary hair slides down his forehead, pricks his left eye, drifts down to the table top, curls into an infinity sign. She follows its progress, looks surprised, then frightened again.

— You got them then. I'm glad, I suppose. Your hair. I thought it was a wig.

— It's not a wig.

— May I?

Her free hand comes towards him. She touches his hair tentatively, first just a finger and thumb testing its texture. Then she pushes all her fingers in, lifts layers of it away, combs it with her fingers, strokes it back away from his forehead, keeps stroking.

— It's like it used to be. Just like. How?

A few hairs drift away, join their infinity friend on the formica. He doesn't care. He can afford them.

— I don't know. It started to grow back after.

— It'll probably all fall out again. Now you're back.

— Maybe. I don't think so. And I don't care if it does.

— Really?

The front door slams. Footsteps in the hall. Vivien looks panicky.

— That'll be Ruthie. There's so much to . . . There's a lot that needs explaining.

— Of course. But I'm going to be honest. You must believe me. I'll tell you everything you want to know.

— Maybe you will. That's not quite what I meant.

The footsteps approach. Vivien's eyes dart around, the kitchen door squeaks open. A bag is dropped heavily to the floor. A cupboard door is opened, slammed shut. Ruthie's weary voice:

— Mum, don't give me grief, I know I didn't call, I was too
— Dad! You're back, I *knew*.

This is the homecoming he had expected. His daughter rushes to him, pulls his hand away from her mother's, wraps his arms around her shoulders, pushes her head against his chest, tries and fails to curl up on his lap. He kisses the top of her head. He squeezes her tight. She nuzzles against him like a cat. Cheaply sexy in a thin gaudy top and a tight glittery skirt. She looks up at him with her old adoring look. This is not the eager fresh-faced medical student-to-be. Yesterday's makeup smeared on her face. Legs that look bruised and used. He pats her on the back. He massages her shoulders. She makes a little whimpering sound.

— I think, says his wife, we should probably visit the hospital.

They get into the car, Vivien in the back, Ruthie allowed her traditional treat of sitting in the front. Howard struggles a little with the gear box, and the women — *his* women? — loyally pretend not to notice. Ruthie is chattering away, filling the space with whatever comes into her head and neither of her parents listen but both are grateful for the noise. Whenever she pauses Howard squeezes her nearest thigh and that is enough of an answer and she starts up again. Vivien opens the gates with the remote control unit and away they go, lurching out of the drive, going more smoothly now along Waymead, towards Hampstead Lane, and the Royal Free Hospital.

— How long has she been sick?

Ruthie stops chattering. She twists her neck to look at her mother. Her mother looks at Howard in the rear-view mirror.

— It was all in my last letter.

— I didn't read it.

— You didn't . . .?

— I'd already decided. It was superstitious maybe but I didn't want to find out anything more. I had this picture, of the two of us together reading that letter in, in our bedroom. Laughing about things. You know?

— It was all there in that letter.

— Everything here was going so well for all of you, it made it hard to come back and disturb things. It also made it easier too in a way. How long has she been in? Is it serious? Where's Simon? At school? Why aren't you at school today, Ruth? You've got exams soon, haven't you?

Howard sees Vivien's hand on Ruthie's shoulder staying whatever their daughter was about to say. His daughter shrugs, looks out of the window. His wife starts to speak, stops, starts again. He notices for maybe the first time in his life how pretty this neighbourhood is.

— There's a lot I have to tell you.

— It's been a long time. I'm glad to be back.

Ruthie giggles. She lights a cigarette. Vivien automatically stabs her window down.

— I am back, aren't I? It feels right.

— There's a lot, says Vivien again, why is she so interminable about this?, that I have to tell you.

Why then won't she tell it? Ruthie inspects her face in the vanity mirror. A combination of spit and tissue and eyeliner restores it to her idea of glamour. She flicks her cigarette out behind her, through Vivien's opened window, watches it bounce and dance down the road, then curls in her seat towards her father, smiles at him, closes her eyes and falls asleep. For the rest of the journey they travel in silence, broken only by Vivien's directions and Howard's mild, dutiful complaints about the traffic.

They walk through hospital corridors. Vivien is shivering. Howard puts his arm around her. She doesn't shake it off so he keeps it there, holds her tighter, and she continues to shiver. She whispers something to Ruthie, who complains, looks at each of her parents, complains some more, then accepts whatever it is that's been said. She pulls Howard's arm away from Vivien and she kisses him.

— I'll be here, she says, almost seductively, I'll be here at the nurses' station when you come out. Don't forget.

She has found a porter to flirt with. A chubby boy with blond hair. Howard watches her go, naked in her needs, and he feels as if there is more he should do to protect her from the injuries of men.

— She really isn't very well.

— Who? Ruthie? What's wrong? What do you mean?

— Not Ruthie, that's a story though in itself. Your mother. I'm sorry. You're going to hate me. You should prepare yourself for a shock.

He doesn't see her at first. He thinks they must have come into the wrong room. He sees four television screens each tuned to the same garish silent station. Four decaying women in four starched beds. A PVC plastic jungle of translucent drip feeds and black earphone cords. The room smells awful. Vomit and shit exaggerated weirdly by a perfume that he associates with his aunt Freda. Vivien waits by the door as he goes in. He doesn't know where to go. None of these carcasses can be his mother. His mother plays bridge for money and tabulates her winnings in a handsome little book. His mother wouldn't allow herself to be seen in a blue hospital nightie. His mother dresses well, elegantly, she always has, carries herself with poise, she is beautiful, dignified, always. She wouldn't have the heart to shrink and waste so publicly. His mother has white hair always neatly combed and clipped, it would never be allowed to grow greasy and stringy and tinted nicotine yellow to echo the smudged hospital walls.

One of the women, the one who isn't watching her TV, starts to roll out a death rattle of a snore. He feels himself begin to panic. He looks to Vivien waiting at the door. She nods, sadly, encouragingly, as if she's been forced against herself to steer him towards a necessary humiliation, a piano recital played from lost memory, a board meeting to which he's forgotten to wear his clothes.

Three of the women are looking at him with suspicion. Their hands wrap protectively around their TV remote control units. He tries his new smile and knows it makes him look mad. He

11

eliminates them one by one, wipes his mind clean of them when they fail the mother test — that face is too broad, that bust too milkmaid large, that body too big. That leaves one and when she is eliminated, he will turn to his wife and she will be pointing at him and laughing: *You see*, she will say, *that's for running away, don't do it again, your face, it was a picture . . . I'm sorry, the joke was a cruel one but you deserved it*. And he will lift his hand towards her and say to her in mock rage, *You got me that time. You, you, so help me, I'LL . . .* And then relief and play-acting will give way to laughter and the two of them will have to hold on to each other, separating now and again to hold their sides and slap their thighs: *I don't believe you did that!* from him. *Your face!* from her, crumpling her own into a ludicrous parody of horror. *God, I don't know*, he'll say, *sometimes you just . . .* And then, finally, after much struggle, her face will settle back into seriousness and she'll look at her watch and say that his mother's bridge afternoon will be over by now, why don't we visit her now, surprise her, we'll pick up some flowers and chocolates along the way . . .

Reluctantly he turns to the final bed, the snoring yellowest woman. This can be done quickly and then it'll be over.

The template for MOTHER holds firm. He compares the elements of this dying woman one by one with the picture in his head and each time he is almost reassured. Yellow jaundiced papery skin, not right, laughable. MOTHER's skin is ivory pale, smooth. Her limbs are rounded, full, not bone-brittle thin. MOTHER's hair is white and combed, not greasy, not unkempt, not like this. This woman is teaching herself to become a corpse. MOTHER is alive. This woman breathes with a rattle. MOTHER would never allow herself to be heard like this. This woman's nightdress has fallen open over wrinkled dried dugs. MOTHER is modest. MOTHER has dignity. This woman is dying. MOTHER is, was, will be, alive.

Howard turns away from the bed. Vivien is pale, tortured by something. He must comfort her.

— This woman is not my mother.

His voice sounds weak to him, quavery. Death has touched him, unmanned him.

— Don't worry. She is not my mother.

12

She is probably somebody's mother, he has seen the wedding ring she wears, poignantly loose on her shrunken finger, and that makes him want to cry. He tries to walk away from the bed. He cannot. He approaches the bed. The woman's eyelids flutter open. Unfocused blue eyes (MOTHER's blue eyes have always been intelligently, ironically focused) wander in their sockets. Even if this woman has taken Dinah's body and destroyed it for her own malevolent purposes she is still not his mother. There is nothing recognisable. The death rattle rolls out, the eyes reach to him, hold, begin slowly to focus. Her right hand waves in the air, rubs the sweat away from her forehead. His mother always used to do that, swipe her brow dry, a nervous habit like a horse swishing its tail which she always claimed not to be aware of. This horrifies him more than anything he has seen, this stupid pattern of neurones and muscles and bone, outside of thought, existing for itself, outside of memory and identity and hope.

— Mother, he says, less bravely than he had intended, sounds spiteful and mean. He hadn't meant it to sound like that.

Her eyes are looking at him now, a vestige of consciousness residing there. He looks away, sees the other three women staring at him. They hate him. They hate this dying woman. Her presence here reflects badly on them. If she has been put in the same room as them then that means, despite the comforting words of their doctors and their relations, that means they are expected to die too. They want someone with them who can fill a bed, who can fill a body. This woman, MO — , he struggles to think the word, has a spirit so tiny, one harsh word could squash it.

She is terrified. She has given up. He sees it. He recognises it. He recognises too the effort she is making, for him, to gather herself into one safe temporary place, to meet him for the last time. He goes towards the bed. He sits on the chair. He grabs her unmoving, unmovable hand and her other hand aims tentatively, clubbingly, towards him. He puts on his new smile and hopes it doesn't add to her fear. She is about to die. She knows it. She wants it.

— Mother.

Her head creaks to face him. She tries to smile back but that tiny vanquished spirit can only twist one corner of her mouth into a leer. Hideously, she is trying to reassure him. He squeezes her hand tighter.

— You are going to be well. You must be strong.

Awful banality. He must say something meaningful. There is nothing meaningful to say. Her hand slaps feebly against the side of his temple, withdraws, slaps again. She leers again. Her eyes show him her horror and her shame.

— Is there something in my — ?

His fingers find nothing strange. Her hand swats him again, and again.

— It's not a wig. It's grown back.

Her hand hits him quite hard this time, rests, drops. It lifts again, remorseless clumsy hand, hits his cheek again. He bows his head, her hand, exhausted now, clubs against his neck and stays there.

They remain like that. He doesn't know how much time has passed, it feels like months, him holding her hand, her other hand resting precariously on the top of his head. Her frantic breathing begins slowly to calm. Her eyes, which have been closed, open again. No matter how long she endures this will be her last moment. Her mouth moves. Gruff sounds tumble out. She is trying to speak. The sounds are ugly, impossible to decipher.

— No. Please. Say that again.

Again the sounds, rubbing harshly against his ear. He closes his eyes, concentrates only on his mother's futile attempt at language. It is a whole sentence, maybe even two. He can tell that by its pitch. But nothing more.

— I'm sorry. I can't . . . You . . .

Her concentration is superhuman. The words come slower this time. Her hand drops from his head, all her energy goes into the final attempt. He has let her down. He should be able to interpret. So what if the dead can't talk to the living. A good son listens to his mother.

She is gasping again, death rattling again. Her eyes close, open. The tiny familiar spirit burrows into a less dangerous place inside this body that has rejected her. A last dying echo

14

of her impossible words, and she turns her head away in self-hating embarrassment, returns to agonised sleep.

— Be strong, he whispers, not wanting the three witches to hear him. Please.

Vivien pulls him gently away. She puts a finger to her mouth to silence the abuse he was about to shout, and leads him out of the room. In the corridor she wipes her hand against his face. He knocks it away, and she shrugs and says, *Oh*, and goes off to pull Ruthie from her romancing position at the nurses' station. It is only much later, when he is lying on the bed, puzzling over the noises he can hear from the room above, that he realises with a terrible, unabsolvable guilt that both women had been trying to stroke tears which he had been unaware of away from his face.

2

HE LIES ON their old marriage bed. The curtains are drawn. It is nighttime but too early to sleep. His shoes have been kicked off. His big toes poke out of twin sock holes. He lies perfectly still, his hands clasped behind his head. Vivien drove the car back from the hospital. The three of them shared a bottle of wine in the living room, and he had tried to answer some of his daughter's questions, until his wife intervened, saying, *Your father has had a long day, there will be time enough for questions, leave him be.* Then Ruthie took a bath and he and Vivien finished off the wine and then she led him to bed, her bed, their bed, lay with him, comforting him a little with the weight of her body. And she had brushed aside all his questions, about Simon, about Ruthie, about Dinah, saying, again, *There will be time enough for questions.* And adding, *Why didn't you read my letter? You should have read my letter.* He had shut his eyes and fooled her into believing he was sleeping, and she had unwrapped herself from him and gone back downstairs.

He lies in darkness. He is home and there is much to do and all he can do tonight is feel sad and half-broken and listen to the sounds that the house makes. There have been visitors down below. He has recognised Leslie's voice, and Charlie's (happy familiarity of his old enemy's voice), and another male voice he can't place, and Vivien has dispatched them all quickly, and he waits for her to return. Below him he can understand all the sounds. Above, though, is different: that could be birds in the loft, or it might be the grumbling of the ancient water tank (wasn't he going to arrange for it to be replaced? he thought he had), or it might be a radio left on, or it might be the sound of light footsteps and a chair squeaking on its hind legs.

16

He had fallen asleep. Vivien's presence in the room wakes him.

— What's that noise? Upstairs. Is it Simon? Shouldn't I see him?

— One revelation at a time, she says, teasing him, skittish. She lies down beside him on the bed.

— What do you mean? he says, and he hears his voice rise up out of a dark solitary hole.

— Which do you want? Touch or words?

Touch. Of course touch. Sex would be yet improper between them. But hugging is good. They hug each other, they put all their tired energy into hugging each other. They fall asleep like that, on the bed, fully clothed, bodies uncomfortably twisted to accommodate each other's unfamiliar shape.

3

THE HOUSE IN the morning. Sunlight filters in through ruched curtains. He climbs painfully off the bed, stretches some of the stiffness out of his body, and arranges the covers over Vivien. She smiles in her sleep. He bends down, pauses, and then for the first time since his return he kisses his wife. It doesn't wake or delight or transform her. Her smile fades, a dream sound drifts from her mouth, might be a gasp, might be the threat of a snore. He tucks the covers tightly around her, waits for the smile to return, and when it doesn't he moves to the wardrobe.

His clothes are still inside. All his suits in polythene dry-cleaning jackets, his shirts pressed, neatly hanging, modestly buttoned, his shoes arranged in soldierly lines with wooden trees inside to protect their shapes against time. In the bottom drawer of the chest, his underwear, folded, vigilant. He pulls off his travelling clothes, bundles them into the corner beside Vivien's dressing table. His nakedness makes him feel alarming. He selects a charcoal suit, purple polka-dotted boxer shorts, thin grey socks, powder-blue shirt with double cuffs, black brogues. He has lost weight since the old days. His thoughts are fast, the crazed urgency of tiredness and adrenaline mixed together. A black belt tightens the waistband and spoils the line and he walks to the door like a spy.

Ruthie is asleep on the landing outside their bedroom. Curled like a child, one thumb suggestively close to her open, paint-smeared mouth. Is she guarding him from harm or making sure he doesn't run away again? He steps over her, a floorboard creaks, one of her hands goes out to touch his bedroom door, the other to protect her eyes. He considers picking her up to carry her to her bedroom. He fears to wake her. He considers

18

lying down to sleep beside her. He kisses her too, he has kisses for everyone, and steps past, along the corridor and into the bathroom, where he cleans his teeth with someone else's brush as he signs his name with urine in the toilet bowl.

Then back along the corridor; he skips lightly over sleeping Ruthie, and goes down the stairs, unlocks the front door, leaves it ajar behind him. He sits on the path. He hugs his shins and rests his chin on his knees. The breeze is nice, welcoming. If he remained like this he would fall asleep again. He stands. He performs recently learned exercises on the driveway. Stretches, twists, jumps, sweating in his suit he jogs on the spot. He feels good. The wild man of the suburbs. His optimism is undiminished. He can do great things. This is his world.

He is being watched. He turns. Bleary-eyed wife and daughter in the doorway pointing at him like he's a zoo animal, possibly dangerous, possibly not, who's found his way into their garden and possibly their hearts. He growls at them, they flinch. He winks, they look at one another. He does a King Kong routine, he's a jungle monster, out of place and time, big and foolish and dignified, able to communicate only through the roar of his voice, the power in his body, the poignancy of his eyes. The women, his women, pretend to be frightened. They shift their feet nervously as they clutch one another in badly acted horror. Until, abruptly, they tumble into marvellous laughter. Howard undoes the buttons on his shirt, he beats on his exposed chest, *HOO, HOO, HOO!* he roars, happy Howard, happy beast, king of the driveway, who consents, finally, to be led by his docile paws into the kitchen for breakfast.

— Where's Simon? It's a school day isn't it?

Munching on toast, sipping orange juice and coffee by turns. His shirt is buttoned, he never liked to wear a tie.

His wife gets up to load the dishwasher. His daughter displays a scholarly fascination for a pizza delivery menu.

— What's the big secret?

His wife returns to the table. She looks around for something to clean, fails, sits down.

— What are your plans?

— Plans? Go to work, re-establish myself, make contact, see

19

how things are, you know? Find out if I've still got a place there. If it's not too late.

— Of course you have.

Dutiful wife refills his coffee cup. Did she use to be like this? He remembers a more casual woman, bored by routine, seldom tempted to play the housewifely part.

— Spend some time with Charlie. Find out what he's been up to while I've been away. I've got the stomach for the fight now, I'm sure.

— Really? And then?

— Lunch at the hospital, see my mother, it was all a shock, I'll be better for it today.

— Can I come? Ruthie trying not so hard to cover the keen-ness in her voice with indifference.

— Haven't you got, school? Or something?

— No. Not really. Can I come?

There are too many secrets here. First Dinah, now Ruthie, who next?

— Where's Simon?

Mother and daughter look at each other. Mother is silently elected spokeswoman.

— In his room.

— Still? He'll be late for school.

Ruthie is looking away. Vivien bows her head. She has run out of domestic tasks to hide behind.

— Won't he?

No one will answer him. Conspirators. His wife shakes her head. So he shakes his.

— I'm getting dressed, says Ruthie, I want to come with you.

— Maybe tomorrow. There's much too much to do today.

Ruthie pouts, is about to argue, then away she goes, one last sulky seductive look back, then away.

Vivien follows him up the stairs to Simon's room.

He knows this door well. It occupies his dreams. Thumb marks on the white painted wood. Empty screw holes for the ceramic plate that used to be there: *Simon*, it said in ornate italic script, surrounded by dog roses and dahlias. The wood is warped, his fists used to beat on it.

— Still?

Vivien looks away.

— I thought . . . You said . . .

How much else has he not been told? Vivien's letters spun out comforting tales, reassuring lies. He read them by candle-light, by gaslight, under electric chandeliers, in the open, by sunlight, sitting on a hill. Ruthie is on her way to medical school. Dinah is flourishing. Vivien runs a gallery, surrounded by friends. Simon plays football for his school team, he excels in all things.

Simon's bedroom door is locked. A radio plays quietly behind it.

4

SHE HAD WRITTEN her letters to him, at least one a month. The letters were chatty and newsy and full of beautiful lies.

— I didn't want you to worry, says Vivien (Is she speaking the truth now? How can he tell?) I didn't want to stop you doing what you needed to do.

They're on the second floor, the trophy room. It was once the nursery, scenes of breath-held terror that their first baby had decided to stop breathing, gay Bugs Bunny surfaces smeared with the stains of blood and mucus and vomit and shit that seemed on fractious sleepless nights to be the only glue that was holding their marriage together. Then out with the cot, roll in a carpet (blue? brown? green?), slap on a new wallpaper and a playroom it became, Victorian rocking horse (piebald: definitely) under the window, a dressing-up box in the corner by the chimney breast, a table-football game against the far wall, there beside it a jukebox and, occupying most of the floor, Howard's joy, the children's reluctant sport, a Scalectrix track. Those things were all put away in the attic a long time ago.

— I could have been cruel, I could have. Told you the truth or made up nasty things. I didn't want to put pressure on you. I wanted you to be happy, not guilty.

— But —

Now the room is a museum of things his company has made. Shelves line the walls filled with plastic products. Unbreakable combs lean against computer casings. Mock-tortoise-shell buttons surround a canteen of translucent cutlery. Coronation badges and frosted tumblers obscure a range of pink Fun-Fones. Transistor radios, table lighters, toy soldiers, portable gramophones, little saxophones, hair brushes and hair clips, wares for the kitchen and table and the outdoor barbecue, everything for the home and office and garden piled high, the higgledy-piggledy history of the family business. The only non-plastic things are in the small corner devoted to Danny: a tray of his drawing pencils, still kept sharp almost fifty years after his death; his comb, breakable but intact; his cut-throat razor. Howard had last looked at these things the day he had discovered the photograph. It was later that day that he had run.

Vivien and Howard sit uncomfortably side by side, he in his suit, she in her bathrobe, on hard rainbow-coloured stacking chairs, sad legacies of LPC's unfortunate lurch into the home furnishings market in the 1970s.

— And I didn't want to give you the satisfaction. I didn't want you to think we couldn't live without you.

Maybe that's the real reason. Maybe they're all the real reasons. He can recognise, by sight, the materials that the things are made from. Melamine and PVC and cellulose acetate and polythene and polystyrene and urea formaldehyde and poly-urethane and polypropylene and acrylic and styrene-acry-lonitrile.

— How much else is there to know?

She smiles. Ironically, a suggestion of cruelty; the possibility hadn't crossed his mind before that maybe she's enjoying this.

— Welcome home.

Howard feels, for reassurance's sake, the damaged photo-graph which he has put in his inside breast pocket. He lowers himself to the floor and plops onto a yellow PVC beach mat-tress. He flattens himself upon it as if he is sunbathing and Vivien comes to lie beside him on their plastic beach.

— I had an affair.

It's said so earnestly, childlike: *I'll tell you a secret, then you tell me one of yours.*

— Who? That architect you wrote about?

— No.

— Charlie?

Vivien laughs at this.

— No. Not Charlie. Would you have thought . . .?

— To get back. At me.

He can hear the anger in her silence. He smiles and rolls

around to face her. He strokes her shoulders. She twists away, lies on her front, rests her chin on a blue puff cushion.

— Things went on that had nothing to do with you. You might find that hard to believe but still.

— I know. But I know Charlie too.

— You know him better than you know me then. He tried. Of course he tried. But not especially. He has to try with every woman otherwise, otherwise I don't know, maybe he thinks he'll die unless he tries — he put his finger up my arse once at a dinner party —

— What!

— But not in a sneaky way, not like Leslie would —

— *Leslie*?

— No I mean. Nothing. We'll talk about that another time. Leave it at this, if you're a woman and Leslie's driving you just make sure you're sitting as far away from the gear stick as you can manage . . . But Charlie does it as a game and because he has to, the secrecy isn't upsetting, it's quite sweet really . . .

— Sweet?

Howard walks his fingers slowly down Vivien's back, feels her flesh through the terrycloth, the road of her spine, the dip at the coccyx, the swell of her arse, which is still in good shape; if he lifted up the robe just a little there would be the penny birthmark on her right buttock.

— It was in the kitchen, I can't remember who else was here, Leslie, Ruthie, Tessa maybe, and I was at the stove and he, Charlie, came up behind me, I knew it was him, smell of cigar, his breathing, I was wearing a skirt, no tights, and suddenly his finger was up between my thighs and up past my knickers and his finger just sort of *wriggled* into my arse and it was all I could do to keep quiet, and it was really very funny and actually quite sexy too . . .

Howard, outraged Howard, outraged that men like Charlie walk around free sticking their fingers up the arses of women, up the arses of *wives*, manages to still the anger he feels; his jealous thumb wants to wriggle between Vivien's thighs, sneak within her buttocks, navigate the channel of her arse, but instead it jerks northwards and settles for the more innocent pleasure of digging in the first few circles of a shoulder massage.

24

— Oh. That . . . But. But no, not Charlie, and not some stupid make-believe architect — that's nice, oh — I don't know where I got the architect from, it used to embarrass me sometimes that the fantasy was so, I don't know, respectable . . .

Both his hands work down her back, smoothing away the tension; down the spine they go, the central road, daring sometimes to steer off into detours — and over there, just a fingertip away, is the ticklish zone above her right hip.

— . . . the truth though I suppose is even worse, suburban — that's, yes, ah — not Charlie, I would never have gone for him, not Leslie . . .

— Who then?

She starts to tell him who it was not and his fingers coax out the names and sometimes when she says a name that doesn't please him he scratches vindictively hard at that unbearable zone.

— . . . and not fantasy John the architect, and not some clumsy teenager like Mark (isn't that what abandoned housewives are meant to do?), not Charlie no, ah, not *Leslie*, not Mark, not John, *oh*, not Charlie, not *Charlie* — Please Howard. Stop it! — Leslie, Charlie, Howard, *Howard*! Miles, Miles, it was *Miles Rodgers* — stop it Howard, I mean it! — Miles Rodgers from across the road.

TWO

HAIR

I

THE PHOTOGRAPH HAD been waiting in a box belonging to his mother. She had asked him to bring it to her, the cardboard box file that had the words Dinah's Things printed neatly on each side. It was where she told him he had stored it, in the corner of his attic next to her old teak radiogram. He carried the box down the stepladder, dented one of its corners on the landing floor, and, for no particular reason that he understood, tore open the masking-tape seal. He removed the photograph from its half-hidden place between rough-edged photo albums and pink diaries clasped with fragile locks. It was a black-and-white shot of his father and his uncle, the inventor and the engineer in workshop clothes, posing for posterity with their very first Universal Unbreakable Combs. The combs were held proudly out towards the camera as if the brothers were a song and dance team who had found a happy substitute for straw hats. Howard had slipped the photo inside his wallet, out of sight. He never did take the box to his mother's house. He left his wife later that day.

It was in the middle of swimwear that he left her. He walked slowly away, through black-and-white American bikinis — one last glance back to see Vivien fingering something green — then on through racks of summer separates and beach shorts and towelling robes, past a sexy-eyed mannequin lounging with a bucket of flip-flops beneath a parasol. Going more quickly now, down the escalator, through the ground-floor halls, leather brief-cases, golden braids of necklaces, a red exit arrow, the sun flash of the glass doors, one last swinging handbag, a final special offer, and, with a successful shimmy past a woman in white who missed smearing his wrist with cologne, he was at the revolving door: wait, nod, hop, push, and out he spun, liberated, a holiday street.

He had not intended this. There had been no plan. The world dizzied him, deafened him. Cars honking. The petrol scent of freedom. There was still time to return to the department store — brush it all away with a smile, an apology, an excuse (toilet? a stomach bug? catch her sympathy that way?) — take her shopping bags, walk on, almost in step, the swimming costume for Ruthie, a new radio for Simon, an unnecessary pair of socks for himself, something in the sale for her. Then the final rush before the meter expired, back home, the secret races through Saturday traffic.

Too much air, he was forgetting how to breathe. He stood near the entrance to an underground station and something sour came up to the back of his throat. A man in a booth sold him cigarettes to kill the taste. He unwrapped the packet, let the breeze tug the cellophane away from his fingers, watched it go, over shoppers' heads. He lit a cigarette and held on tight to the nicotine high. His left hand scratched at the shiny bald patch on his crown which he still, nostalgically, thought of as a spot. He went with the crowd.

She would be walking out of swimwear, hunting for him, his name whispering uselessly through her head — then the pang of nauseating uncertainty would hit her, a rush of nerves almost to panic; she would stop, stand still, gulp for control with big efficient breaths, compose features, dab hair, calm herself by looking at something fixed, an elevator door, a red exit arrow, a mannequin under a parasol, then a little step in one direction and a bigger step back, the brisk return to swimwear, deciding, nodding, yes, they would find each other better if she stayed where she was. And the shoppers would go home, and the display cabinets would be emptied, and the window lights would be killed, and shop assistants would touch up their hair and fix their makeup and put on their overcoats and she would still be there, Ruthie's new swimming costume abandoned near her feet, blank-faced, maybe hating him, suddenly worrying for him (was he hurt? sick? murdered? it happens) and then worrying for herself, a rush of self-pity, then defiance, some of the time stubbornly trying to think this thing through, make it normal somehow, some of the time knowing he had run, gone, absolute, leaving her with a man-sized hole for company; until, finally, a polite distant touch on her arm, a shop assistant made unfamiliar by an overcoat, a murmur, courteous yes but entirely relentless, *We are closing now*, a pause to allow the words to find their inevitable resolution, *madam*. And no matter what happened now nothing could be the same again.

He stamped the cigarette out on the pavement. Nicotine and hope, nicotine and fear, the combinations made him giddy. His forehead was wet from the cold pressure of an unknowable future. He was near another department store. It might even be the same one, he couldn't tell. They might see each other in the street outside. Man and wife, a destined return, the sign of a necessary truth he was powerless against. He would say he had needed some fresh air and she would smile — *Oh, I thought . . . it doesn't matter* — and the cigarette in his hand would provide the acceptable focus for all the horror she had felt. He lit another cigarette. He moved towards the door that was spinning shoppers out into the street. He stopped, shut his eyes; someone hit him on the shoulder and that knocked him swivelling half-around but he managed to stay upright and blind. If the first person he saw when he opened his eyes was a man then this thing would be over. In front of him a tiny old woman with a silk scarf over her head and a jacketed dog scurrying inside her arms was trying to catch a taxi driver's attention. Howard the tyrant pushed in front of her. He stole the cab and told the driver to take him to Paddington because there, long ago, was a seedy hotel in which his father had died with a red-haired girl.

THE SUCCESSION OF patriarchs. Howard the begetter of Simon. Daniel the begetter of Howard. Jack the begetter of Daniel. And before him, nobody. A shtetl darkness, no names. In the beginning was Yakov Levy. After he bought his ticket to New York he changed his name to Jacob. Shortly after he married he became Jack Levy. There would have been others before him, he did not give birth to himself, and there was some family talk about rabbis and merchants and tailors, but Jack Levy died in 1938 and whoever his begetters had been, they were not as interesting to talk about as horses and cards and politics.

Jacob came over to England on a boat. Three days' heavy sailing across an ugly North Sea from Hamburg. Steerage sweat and spray on his second-best suit. A brown suitcase in his rural hand. In his suitcase were the proceeds from the sale of two Galician cows, his family fortune (this is a story many times told — after a generation of its telling the fortune had grown to magnificent proportions: a king's ransom, a workshop's mortgage, a Rothschildian dowry). A crush to leave the boat. New York awaits (he had bought the ticket from a fast-talking Salzburg merchant with a good grasp of Yiddish and character). Ahead, the dockside was filled with New York gentry — or poor recent immigrants and petty London conmen who passed the time in between boats in the gin shop overlooking the landing stage. A bustle, a crush, a wild storm of gentry, including the man with the roguish brown hat and the green sharper's suit and the little tuft of a moustache who greeted Jacob with a kiss. *Welcome*, in Yiddish, *my landsman, we are friends*. Jacob broke the embrace. He looked around and breathed in the rich air of the promised land. *This is New York?* His new friend grabbed his arm. *Of course it's New York. And keep this quiet, we must talk in whispers, oh jealousy, you see that man with the badge? he's a tax man, government man, a vulture, we must talk in whispers, the vulture, you are lucky to be recognised, you are from . . .?* Jacob considered lying. He shrugged. *Ostrov.* The little man clasped ahold of his hat as if the sound of the name was an angel's song from paradise. *And me too! Also from Ostrov, You are like a brother, let us drink, let us eat, tell me all the latest news, my heart is sick for knowledge of Ostrov, and your family name?* Jacob told it, angry at himself for his mild suspicion of this amiable friend. *The Levys of Ostrov, such a notable family, your pretty sister is she well? Her eyes, a man would commit crimes for her eyes. And your good father? His business is flourishing? My name is Lenny. Lenny Leverton. No? No recollection? We lived just outside Ostrov, my father the butcher, you know him? No matter, we were poor, but good. And I who have lived here now in New York for five years, still every week I come to the landing place hoping for news of them. I used to send them money, but no longer, I haven't heard news of my family for so long, the bastards, the peasants, the Jew-haters, you are a lucky man to be free of them. Here in New York the streets are paved with gold and I shall take you for a ride in a Broadway taxi all the way up and down Coney Island to Fifth Avenue. Come. Let us walk. Let me carry your bag. Let me kiss you on your honest cheek. No, it is no burden. You must be tired after your voyage. Happy day! A blessing to the Almighty One who has effected this reunion! Quick, the vulture has spotted you. We must walk faster.*

The vulture, the little man with the metal badge pinned to his overcoat, waved at Jacob Levy, shouted to him over embracing couples, through solitary men sitting on their suitcases, jumped up and down, shouted again, through the arms of sailors in uniforms and shipline men, but his words were lost in the air and the hubbub. Lenny held Jacob's suitcase in one hand, linked arms with his new friend with his free arm, pulled him swiftly away from the port, into narrow tenement streets, around corners, briskly, too briskly for tired happy Jacob to take in any of it (and a sense of direction has never been one of the family gifts), a maze of tenement streets, rights and lefts, through archways, sudden sights of water — the ocean, Lenny Leverton would inform him, the Pacific Ocean — then snaking back away from it.

And, outside a narrow tall house, *This is your new home, not a palace, but clean, respectable, an honest house. First you must wait a moment outside, I have lived so long without hope, a shell of a man, so long it is since I met a landsman, that the room I had ready has fallen into some misuse — no you are too generous, it matters to me, that is the important thing, everything must be good and right and spick and span, just have a moment's more patience, you are a saint of a man, food will soon be on the table, a bed with sheets of linen and lace will be yours, and work, I will find you work — are you a tailor? a worker with fur perhaps? — but first I must organise your room and tell my wife of the good news, wait here, just a moment longer. A blessing upon you. I will take up your bag.*

And Lenny Leverton opened the door with his key and shut it quickly behind, and Jacob waited for his new friend to return and let him in. And he waited and he waited, on an empty narrow street, in front of a tall narrow house that was missing some of its window panes. And finally tiredness and hunger overcame horror of appearing rude, and he knocked on the door, and he knocked and he knocked. Until eventually, perhaps an hour had gone by, a large red-faced yellow-haired woman lifted open a first storey window and shouted at him in strange angry words, and when he stayed stubbornly where he was, she started to throw pieces of brick at his head. Still he waited, with a broken head, with blood running down his face through the grime and the sweat, and also tears. Until two men in uniforms came and battled for his future in words he didn't know, and stood on either side of him and lifted him off the ground and carried him easily for maybe a mile and deposited him in front of a shabby large house on a shabby road, where the vulture, the little man with the glinting metal badge, looked at him sadly and shook his head, and led him slowly to a narrow cot in a shared dormitory room.

He remained there for two weeks in the Poor Jews' Temporary Shelter, drinking beer and schnapps with a hard man from

Warsaw who hated everything. He was thrown out after those two weeks because that was the limit of his stay, but the little man who wasn't a vulture at all took pity on Jacob's misery by providing him with a large bag of unmatched buttons to sell on the Commercial Road.

He was not a natural salesman. Not Jacob. There was a diffidence about him, a sadness. His sorrow grew from inside until it entirely surrounded him, and even if he had been a better button seller, still that bubble of sorrow that surrounded him would have scared the customers away. He would arrive at the same pitch each morning, to stand in front of a kosher butcher's shop, between an old man who sold brass musical instruments and a young boy who sang happy songs in a sad beautiful way, and Jacob would just whisper hopelessly the words the non-vulture had taught him, *Buttons. I have buttons. Who will buy my beautiful buttons?* He kept his attention on the men in the crowd, hoping to find his enemy Lenny Leverton, the ruiner of his fortunes, and his words would be lost in the pure sound of the young boy's harmonious voice, unbroken songs of joy that had so sad an effect, and the blasting of brass instruments, and the squabbles of wives with the butcher man.

Among his few customers was a girl, the singer's sister, a tiny girl with a pretty face and nervous manners. Her name was Sarah. Her last name was Bremen because that was where her father, a master tailor, had bought his shipping ticket. Something about Jacob attracted this girl. It might have been his sadness, or his diffidence, or the disappointment inside him, or maybe it was his strong short body, his unschooled muscles, his placid head. Whatever it was, she bought at least one button every day, and after a week of this they found themselves taking walks down the Whitechapel Road after dark, and sometimes, he didn't know quite how it had come about, he would discover her hand tucked into his.

Sarah learned the story of the dock lounger. Jacob learned to overcome both his timidity and hers. Before he had reached the bottom of his bag of unmatched buttons, Jacob was offered a job as an assistant presser in Mr Bremen's workshop. He proved to be a good steady worker. He was given a room in his employer's house. Sarah and he got married. They moved to an apartment across the hall. They raised five children and they buried two others. Sarah remained shy and pretty and frightened. She never learned to read or write and her husband never saw her body naked in forty years of marriage.

Jacob improved his name to Jack. He was good at his job, a valuable worker, reliable, unambitious except to win at horses and at cards. Sarah was frugal with money, didn't complain too much, she cooked fairly well, she kept their flat clean. Each of their children had a permanent epithet: Rebecca the beauty, Freda the frail, Bertha and Sam (twins), she the mischievous one and he the fighter, and the boy Daniel, the idiot.

Jack's unschooled muscles became surrounded by a plump layer of almost prosperous fat. His placid face grew rounder and bigger. His hair retreated. There were only two things that could disturb his placidity. One was the playing of cards, and then he would sweat, he would grin, he would drink beer and schnapps, he would curse beneath his breath, or whisper half-remembered Ostrov blessings. The other thing to disturb his placidity was any mention of Lenny Leverton, the dock lounger.

Dinah took Howard to the East End when he was thirteen, a week away from his bar mitzvah. She showed him where his father was born, and she showed him where the Cambridge Picture Palace used to be. That, she said, was Schein's Gentleman's Hairdressing Saloon, that was Strongwater's delicatessen, and that was where the intellectuals drank lemon tea at Goide's restaurant. There were still Jews there, bearded cowboys in wide hats and heavy black coats, their wives' bodies sturdy beneath wiry brown cakes of wigs, the boys with heavy glasses and wisps of hair in front of their ears — some peculiar covenant with a missing God (why are they so obsessed with hair? Did their God get ripped off one day by a celestial hairdresser?) — the girls inside party dresses and white tights to hide the mystery of their legs. But they were soon to be outnumbered by the Asians, a bright red mosque would be built where the synagogue used to be, no longer the smell of fried fish to sour the air, but biryanis and tandoori meats and vindaloos. The tenements were destroyed by the Luftwaffe and the Council, the hospital is still there.

Danny, like the other children, drank in the tale of Lenny Leverton with his mother's milk. It became a kind of inspiration to him, and when Danny became rich, sometimes he would say in respect of his growing fortune that he owed it all to Lenny. Long-nosed Danny. Maladroit Danny. Daniel who can't dance. Daniel who has the gift of healing machines and making them grow. Loveless Danny.

Imagine the flat. Make believe the years and the deaths never happened; sniff some alien London air (is that the smell of stale chicken soup on the landing or just dried-up piss marbled into the wooden stairs?): step inside.

It's night. The family Levy is asleep, or at least should be. In the parlour Sarah shares a single bed with Rebecca while over on the sofa Bertha whispers sinister ghost stories to keep her sister Freda from sleep. Next door, in what was once the marriage room, snoring Jack shares the marriage bed with bulky Sam; but the cot against the wall is empty.

Ten-year-old Daniel, in his string vest and good trousers, sits on the high kitchen stool at the kitchen table. His small narrow body is hunched forward. His black hair falls over his forehead. In a few years Daniel will develop a liking for Rudy Vallee, and only partly because his sisters, preferring to translate his Jewishness into Italianness, often declare a resemblance. But Daniel's only ten now. He hasn't heard of Rudy Vallee, he doesn't even know he likes music. And this ten-year-old Daniel is performing a forbidden act. Daniel has destroyed five wireless sets already but there's a compulsion inside him, the compulsion of idiocy, or genius perhaps, if that's what he is, and the word isn't just the desperate expression of mother-love towards a misfit boy with a compulsion to fiddle. Sometimes he reaches with his left hand towards the open jar of his mother's home-made haimeshe pickles in front of him, slowly dips into it, a shy shark of a hand, pulls out the pickle, shakes off some of the excess brine, draws it circuitously towards his mouth, avoiding the airspace over the dismantled wireless in front of him, chews upon it while he works with screwdriver, by candlelight, to understand the mysteries of this machine, which sits in its different parts surrounded by shallow pools of brine.

A noise. A succession of noises. A sharp sound, hands clapping, then someone's breathing and a sisterly laugh. Daniel jerks round, the toe of a half-eaten pickled cucumber protrudes from his mouth.

— You've been forbidden.

Rebecca, the beauty by candlelight, stands half in shadows, half out. There's something strange about her appearance. Daniel struggles to gulp down the rest of the pickle.

— You won't tell?

— Why won't I?

— What's wrong with your face?

— There's nothing wrong with my face. There's some who admire this face. You should be in bed, Einstein.

— Why have you got your coat on? Are you going out?

— And they call you an idiot? It takes a genius to make such a deduction.

— Where are you going? What's that on your face?

— Let's make a deal.

— What kind of deal?

When Danny told this story to Dinah his worn, crazy face took on the expression of the ten-year-old he had been. When Dinah passed the story on to Howard she had tried too hard to make her son's face into the image of her dead husband's. The suspicious child still fixated on the lipstick spoiling his sister's face.

— Our secret. You didn't see me, I didn't see you. We're both fast asleep aren't we?

— I suppose.

— You had better get that wireless put together by the time the ancients are up.

— I've worked it all out now. Where are you going?

— People who don't ask questions don't get told lies.

Rebecca flashes her brother her second-best movie starlet's smile, holds it, checks it in the mirror, fixes something astray in her hair. By the time she approaches the door Daniel is already concentrating on his radio. It wasn't quite true what he said, there are still a few things to sort out before he can put the radio back together. His steady hand reaches out for the final pickle in the jar.

The idiot. Danny. Victim of cataclysm. Founder of fortunes.

Beloved of memory. Late the next morning, a sabbath morning, Jack Levy goes into the parlour (magically transformed from its nighttime bedroom for girls). He's going to switch on the radio to hear the big race. Fat little Jack doesn't drink, he doesn't run around with women, but he loves to hazard money. Friday-night card games at Solly's place that stretch to Monday morning. The horses, of course — courtesy of a bookie's runner who's never been seen to run, just lounges like Lenny Leverton on the corner of Brady Street and Whitechapel Road, smokes a cheroot and tips his gentile hat to every woman who passes by. He'll even bet on the neighbours. Jack Levy and Mr Singer from across the landing will stake money on arrival times of local babies, on the percentage of Jews to Irish in the tenement, on how many pimples can be counted on the face of the travelling chazan, on raindrops racing down rackety windows. One day Freda saw them on the Commercial Road counting the buses going past, Mr Singer's face already turning a greyer shade of grey. Her father shooed her away, but later that evening they ate very well. (She told this story to Howard and Charlie in a quiet afternoon at the factory. It was meant to be a warning against gambling. It filled them both with the desire to risk and to win.)

Jack switches the radio on, settles into his armchair, a copy of the *Daily Worker* ready to be fanned across his belly. The light winks on and glows a tiny orange. The radio warms. Live voices slowly rise up through drowning crackle. A dance of lines waltzes down Jack's big moon face and ends in a scowl. This is not the big race. This is not even English he is hearing, these light voices rising musically for unfamiliar reasons. Jacob wheezes halfway out of his chair. He bellows the idiot's name — *Danny!*

But Danny isn't there. Danny's underground, riding a tube train, knees close together, orbiting the city in anonymous safety, his head filled with images of valves and levers and cogs and gears and motors, as he constructs perfect machines in his imagination.

The rest of the family come to the radio. Sam, cigarette in his mouth, cap on his head, shirt sleeves rolled up to show off his surprising muscles, is the boldest one. He adjusts the tuning.

Rolls the dial into different worlds one after another. It is a miracle, what the radio has become. Five times as strong as before — and the BBC still there if you looked for it. But far-away places hustled now for attention. Warsaw, Paris, Rome, Oslo, even New York, the family's spiritual home. And Daniel, for a while, before he damaged the radio again, this time beyond repair, became the family's hero.

Danny's body doesn't grow. He's the horror of his mother. The same skinny runt of a genius-idiot. The same flop of black hair bouncing off his greasy forehead. Only his nose grows, a caricature of a Jew, which his manner and dress live up to, dark-clothed, dark-complexioned, head bowed low against the insults of the Irish. But Danny has a gift, the ability to tamper with machines and make them grow. He offers out his expertise for money, he doesn't know quite what he's doing, and the money doesn't interest him — most of it goes to his mother for cigarettes for Jack and Sam — but he cherishes the blessing of new machines to play with. Mrs Roth's sewing machine, the presser that Jack operates at Mr Bremen's workshop, the Morris car that belongs to the rent collector, the projector at the Cambridge, where the girls like to go to flirt with the neighbourhood boys while Fred Astaire dances. And in between his machine tasks, Danny piles up his school books, puts them nonchalantly in his satchel, and creeps away down to Aldgate tube to ride the Inner Circle of the Metropolitan and District line and build things in his head.

Bertha got a job in the cigarette factory. Frail Freda who outlived them all became a secretary in a jewellery business on Regent Street. Sam, the drinking Jew, the fighting Jew, worked as a signwriter's apprentice. Rebecca married a scholarly school-teacher called Sidney and before the gossip of her adultery drove them to Australia they lived in a nice house in Stamford Hill where they held meetings of the Left Book Club and talked about Spain. Danny left school behind him for ever and went to work in a plastics factory, and that's where he invented his first machine, which moulded buttons better and faster than any machine or man had ever done before.

He is the creator of the family fortunes, or — as Jack would have it — its repairer. Danny began his own company, below a sweatshop on Hanbury Street. He planted machines on the workshop floor and watered them with his genius. Sam became his partner, the dictator of the factory. The parents were sent off to a large house in Stamford Hill which they never learned to feel comfortable in. Danny lived for his machines. Until one day his curly brilliant head was turned by love.

Her hair was red and cut short like a boy's. Round face, snub nose, green eyes, oddly large ears. In a short black dress and jacket with poorly made mock-ivory cellulose acetate buttons, she was sitting opposite him one day in a tube carriage on the Inner Circle. He had kept to his childish habit of riding the train to think, orbiting the city of London on safe certain tracks. He was sitting there, in a black suit and an open white shirt, using a tongue-licked pencil stub to draw shrewd machine lines on the back of a torn brown envelope on his lap. He looked up for a moment and he saw this girl, and she was laughing at him. He must have blushed. Looked back down to the envelope, his head suddenly empty of everything but the girl. The lines, which a moment ago represented the beginning of a blueprint for a hairbrush mould, meant nothing. He tried not to look up again. He was worried that she was still laughing at him. He looked up again. The awful disappointment that her eyes were no longer on him. Daniel tried to make sense of a hairbrush machine.

On his second orbit, he looked up again from the plans of a machine that would never be built. She was leaning forward, chin on fists, elbows on knees, legs daringly uncrossed. Her eyes, puzzling and frightening, were fixed a little sadly on his. *Hello*, she said. *Yes*, he said, imagination full of her, sense of propriety confounded by her alertness and presumption. *My name's Ruth. What's yours? Daniel. Danny. Daniel. I'm thirty-two*, he added, not knowing quite why, *and I'm my own boss. I own a factory where I make things that make things*. She laughed at that and he laughed too even though she was making fun of him, but he was glad that he had the capacity to make her laugh. He would have tied tin cans to his limbs and performed a crazy tumbling dance if that would have pleased her. He would rhythmically slap the red cheeks of the fishwife sitting beside the green-eyed girl. He would go up to the bowler-hatted gentle-

41

man at the end of the aisle and lift up his briefcase and empty it over his bowler-hatted head. He would insult his father, he would curse his mother, he would pour blessings on the name of Lenny Leverton, if that was what she required of him. *And*, he said, *I'm not married or even about to be. I care nothing for politics, I don't understand war, and I have some regard for beauty but — forgive me for this speech, never before did I talk like this, a blessing on the name of Lenny Leverton, may all his children be prosperous — I have never seen anything in this world nor imagined anything in the one above to compare with the beauty of your eyes and the redness of your hair. Excuse me. I am covered with confusion. This must be my stop. You might tell me where you live?*

She didn't tell him where she lived. She only stared at him, grinning behind one hand raised to cover her mocking mouth. Daniel Levy sadly removed himself from the train. He stood on the platform, no idea where he was, and stared at a bright poster advertising the countryside papered onto the wall. He looked down at the tracks. He examined the electrified rail, considered its engineering, and he might have vaguely contemplated suicide. Then he began to calculate. He was good at calculating. He waited for the train he'd been on to circle back to him. He checked the windows of every carriage of every train that came through, just in case he had made a mistake, and when, fifty-seven minutes later, he re-entered the carriage he had been sitting in there was no Ruth, no green-eyed girl.

He became like an old man. Touched for that moment by love, then untouched. He was thirty-two, unmarried, with crazy eyes and a long nose and a flop of curly black hair. In between his daily orbits on the Inner Circle he spent his time on his factory floor. The Levy Plastics Company now filled the building on Hanbury Street, machine odours battled with the smell of beer from Charrington's brewery. Danny's Universal Unbreakable Combs were made there, the capacity to build them slower than the industry's demand to receive them, but Danny no longer helped out with that. It was left to Sam, his brother, the foreman, to marshal the forces of the Irish teams to build those big grey injection moulding machines. Instead, Danny, inconsolable, made a corner for himself in the mousy shadows of the exit door where he and a Slav who had no English built new machines, ugly squat monsters that looked malevolent, made for unimaginable purposes.

Sam drank the nights away in green-painted pubs singing sentimental songs in English and Gaelic, and returning home, if at all, often with a broken head. But Sam one day got married. To a nice Jewish girl called Esther, who removed some of the fire from him, and replaced it with contentment and father-love and good-food fat. And this paragon of a wife had a younger sister who was called Dinah and who looked, her admirers used to say, the spitting image of the singing dancing star Deanna Durbin.

Danny and Dinah, the machine magician and the shy girl of the dancing Hollywood looks. Sam and Esther, with Jack and Sarah's connivance, did everything to throw this perfect pair together. Picnics in Springfield Park. Weekend jaunts to Westcliffe-on-Sea. Trips to the cinema. High tea in the parlour, where the two would somehow always end up alone.

Daniel frightened Dinah. She was scared of the intensity of his look, how he would walk up and down like, in Sam's words, a lost fart, of the way his cup would splash tea onto his saucer and knee and he wouldn't care, just carry on staring at her like she disappointed him utterly, his leg continuing to shake up and down, the only outward demonstration of his sadness. And she was scared of the odd black and ginger stubble that was beginning to beard his face. She was scared of his absorption in a private mystery that he would never reveal anything of, no matter how much she prodded and darted with quick canny questions. And she was scared of Daniel's dark unfinished machines that cluttered the mousy corner of the factory. And she was scared too, pretty Dinah, of the way his sad ugly face would fill her dreams at night.

Dinah grew in confidence, the more time Daniel spent in her company. He would take her at night to the new factory, which was being built on a wasteland in the west of London. He'd show her the new machines, but never explain their purpose. They had names, these machines, girls' names (but none with her own), the Rosie machine, modest and little, a grey steel little workhorse with a shuddering arm like a little mutant sewing machine, the Agnes machine, gaudy and ornate, black steel studded with fragments of coloured glass, a flirtatiously weaving metallic motion; the Bertha machine, solemnly pretty, built for his mischievous sister, dead of TB; and then there was the Ruth machine, painted red, short and slim, with a bulb at the top a lot like a head, and a red button at its groin which never was pressed. Daniel would take Dinah to visit his machines, and in return she would take him swimming at night in the ponds on Hampstead Heath.

This difficult romance ended in a difficult marriage. They wedded, Daniel and Dinah, in a small dusty synagogue on the corner of Brick Lane, where Sarah and the other women wept in the gallery and Rebbeca looked Hollywood-beautiful in a modern yellow dress (and the women broke from their weeping to make nasty gossip about Rebecca's wifely vows). Daniel and Dinah lived in the house on Stamford Hill. Daniel continued after his marriage to ride the Inner Circle and construct peculiar machines with girls' names, while his unbreakable plastic combs and buttons were sold around the world. They had one child, a quiet boy named Howard, declared by Sarah to be the reincarnation of her lost husband Jack whose lungs had finally coughed themselves to pieces. (A declaration weakened only a little by the identical one she had made a year earlier about chubby little Lawrence, Sam and Esther's proud progeny.)

Daniel died, scandalously or joyfully, whichever way you look at it, in a cataclysm of fire in a rented room near Pimlico. His son was a year old, his business, the Levy Plastics Company, was thriving, and he had found his Ruth again, or a simulacrum of her, maybe conjured up with engineering. Nobody ever reported seeing them together. But he had a woman, he told his brother Sam briefly about his secret love, a young Christian girl with fiery red hair. (And in the way of these things, Sam told Esther and Esther told Dinah and Dinah kept it to herself but then, one tipsy night on the eve of his own marriage, told Howard.) Danny met his woman once or twice a week in Pimlico. Nobody can know what went on inside that room. We do know though that it ended in fire, a leaking gas pipe perhaps, a careless cigarette, a foam cushion, the building aflame, everyone evacuated apart from two. They were a man and a woman, the landlady said, a little Jew and a girl with red hair. But when the firemen finally doused the flames and axed their way into the gutted rented room, all they found were two human bodies, scorched beyond recognition, melted together at the hip and mouth.

3

HOWARD WAS NAKED. He sat, smiling, comfortably hunched, on a pink polyurethane armchair at the Hotel Continental. The room was a cocoon and the cocoon's walls were beige. Floral curtains hid it from the street and the moon. On the wall beside him the painting of a sad child permanently leaked a big round tear from one of its big clown eyes. A metal wastepaper bin stencilled blurrily with red and white roses was waiting between his feet. In front of him, piled up on the coffee table, illuminated by the overhead light, was the rubble of the things that he owned.

Howard was making decisions. He was killing a dead man.

Too much evidence. Throw it all away and identify the corpse beneath. Credit card slips, department store receipts, restaurant bills. Who was this paper collector? What moved him? Maybe he had held on to all this paper in the good citizen's expectation that someone with a uniform or a toolkit might one day demand them from him. No one ever had but still he had done it, the dry determination of passing events to establish themselves as habits. All the *Thank You For Shopping At*s and the *Cardholder Please Retain This Copy*s and the *This is NOT a VAT receipt*s. Had he actually *wanted* any of those meals?

Driving licence. No penalty points, address correct, he had better keep it. One day he might need to pretend to be himself. Health food shop receipt (*Please Debit My Account By £27.43 Thank You For Shopping At Solistic*), the goods itemised, pulses and beans and muesli and homeopathic remedies, most of which would have entered his mouth or been smeared across his body, few of which he would have ever expressed a inclination towards. Accident. Habit. Only.

Howard examined his predecessor's life. Howard Levy. Forty-seven years old. Family man. Consumer. Businessman. Howard Levy's business cards were made of stiff clear acetate etched with black ink. They looked like this:

L P C

FOUNDED 1943

'the future is plastic'

Howard A. Levy
Managing Director

1–9 Western Avenue Estate London W3   0181 947 1818

He folded them in half and bent them into quarters and let them fall, buried in roses.

Once before he had run. It was a few years ago, after Charlie had insisted on taking him to his club in Soho. Howard had escaped while Charlie was occupied at the bar charming over some girls called Alison. Afterwards, he had been walking through Soho alone and a drunken wisp of freedom had made him shiver and he had resolved to go. He just didn't know where. So, after sitting through the second half and then the first of a pornographic movie called *My Hole Dreams*, Howard had spun slowly through West End streets, looking at the lives of all the people he saw, and wanting, humbly, to join them.

There had been a suburban couple, middle-aged, probably worked at the same office, probably married but not to each other, who were leaving a theatre at the end of the show. They were hungry. They needed a restaurant now. Howard had watched them examine the window of a steak house, red velvet banquettes, green lighting, over-priced comfort. He glanced at

47

her. She shrugged. He tried a hand on her shoulder. She accepted it. Together they pointed at items on the menu, agreed on the reasonableness of the prices. Howard the novice runner had stood beside them, unnoticed, perhaps invisible. He would have liked to be their guide, to become the angel of their romance. But how to do it? Just intervene, lay a hand on the lovers' backs, introduce himself with an air of cosmopolitan grace, force himself not to be repelled by their repulsion, the way they would shrink from him, the city madman; and what would he say? Don't eat here, the food will be shit in here, you're a street away from Chinatown, walk down there, go to a restaurant that worries you because it's strange, smile at the staff, sit down at a brightly lit table, consummate your affair with foreign food. Howard had cleared his throat. The couple passed between him and the window. He had started to speak, unnoticed. They entered the steak house and primly separated when the bow-tied waiter approached. Not bold enough. He had told himself to be bold.

An old gent in a camel-hair coat, sharp little man, perfumed silver hair, was leading a party of other sharp little men towards Leicester Square. *Over there*, he said, making them stop right next to Howard. *You see it?* He pointed up at a first-floor Italian restaurant. *That's where the Windmill used to be, where the girls wore no clothes.* The party of sharp little men looked up at the building, dutiful party who had already been told too much of what used to be where and what was bombed in the War. *Right there*, the old gent said, *that used to be the Windmill.*

Howard had wanted the old gent to be his friend. He had wanted to be sharp enough and small enough to join his party. It couldn't be an accident that the group had stopped right next to him. Howard spoke.

— It didn't.

Nothing happened. Howard said his line again.

— It didn't.

The old gent swivelled around, a fairground boxer ready for his next opponent.

— Are you talking to me?

— I said it wasn't. The Windmill was never there.

It gave Howard no pleasure to contradict the camel-hair gent.
It just seemed like the surest way in.

— I beg your pardon?

— It was on the next street over. The one parallel. That's
where the Windmill used to be.

Camel-hair gent looked at Howard. His party of sharp little
men looked at Howard. Howard had misjudged this. Howard's
lie had threatened camel-hair gent's authority. The gent would
have to assert his authority and Howard would not ever be his
friend.

— Fuck off, said camel-hair gent. The party had moved
briskly on.

There had been no way in. A chubby girl sitting by herself
in a coffee house with a cup of cappuccino and an annotated
copy of a paperback play scared Howard off when he
approached her because her look clearly marked him as a sex
criminal rather than the theatrical impresario he was about to
be. The dapper crowd of Japanese businessmen tolerated him
only briefly before it became clear that he was the controller of
no prostitutes. The ancient sisters elegant in evening dresses
ignored the cab he hailed for them and stepped instead into the
one behind. The waiter in the brightly lit Chinatown restaurant
would talk only about subjects referred to on the menu and
even then not for very long. A madman had reared up at
Howard on Little Newport Street and asked him if he was
enjoying the game but Howard had no envy for his life and
made no attempt to enter it.

He had returned home to his interrupted life (*You're late,
darling,* his wife had murmured as he climbed into bed. *How
was your evening? How was Charlie? Were things all right
between you for once?*) He held to himself the secret of the
interruption and that had been enough for a while.

And now. This time he would do it right. These things he
discarded with a smile, a wince, a sneer, rolled into rustling
balls, broken in pieces, dropped like dead leaves into the bin
between his feet: invoices, membership cards, receipts, a food
shopping list and a list of things to do. Crispy combustible
paper.

Business cards. The marketing managers and sales managers and managing directors he had worked with or against or above. Andy Astaire from Consolidated, Donny Byrne from The Extruder People, Robert Crawley from International Polymers, Jon Dover from Dover Soul Productions, and all the rest of the Feldmans and Golds and Hughses, the Combs 'n' Cuts and Financial Services and Hair & Nows. There was the card from the music shop where he had bought Ruthie a piano accordion that had received one lesson and then disappeared wheezing into the attic. The card that he had nostalgically kept or neglected to throw away of the place in Provence where the eating was outdoors under trees and the chef did pretty things with courgette flowers. The Japanese place in Swiss Cottage. The Indian in Bethnal Green. The French in Hampstead. In they went, a kindling of cards.

His legs and buttocks kept sticking to the chair. As he sorted through the rubble, grinning, assassinating (there's a glamorous one, *Signature Client, Merci*), he would lift each thigh in turn away from the plastic cushion to enjoy the sensation of gentle unpeeling. His life had been accident. Precedent and habit and other people's choices which presumably, in their turn, were equally determined by place and circumstance and *their* other people. Membership cards. So many, too many, watch them go. There's the golf club that Charlie had suckered him into joining and which he'd never attended — he had heard talk that Charlie received his membership free if he got five others to join. Rip it in two, let it fall. There went his friendship cards with various museums and galleries (purchased by Vivien, equally friendly to unresisting spouse). The card with the details of his blood (A rhesus negative) and the expression of his benevolent desire to donate after death any organs to whoever might be going without. Had he ever had that desire? Truly? He kept the credit cards.

He was killing himself. It panicked him. It exhilarated him. His penis teetering on the smooth cliff of the seat cushion shrank, and that might have been from the cold or it might have been the fear. His calf muscles kept twitching. The hairs on his legs trembled from an unaccountable breeze. He was left with photographs, £347.32 in notes and coins (most of which had been destined for a motor mechanic who only took cash), a brown hide wallet with brass corner reinforcements, half a pack of cigarettes, two near-empty boxes of matches, and the untidy pile of his clothes lying in a sad bundle like a roadside kill. He pulled a white monogrammed handkerchief and a company tie out of one jacket side pocket and two sets of keys (house, car) from the other. He dropped them into the bin.

The floral patterns on the curtains that hid him were growing more distinct. It was already daylight. London was wakening. Soon everyone would be at the funeral. The empty gap by the graveside meant for him to fill. There would be gossip during the service, clucking malicious tongues, Vivien's panic during the walk from hall to grave, harsh words during the laying of earth, sympathy-surrounded Vivien — sympathy would be the worst; he saw his wife wincing with every kind word, he tasted her grief, her unbearable confusion, he could raise a hand to the empty space beside him and feel her heart flailing with the sudden push into the unknown.

Is that why he had run? A funeral? Hardly. He had seen dead people covered in earth before. He spent a fruitless time, sitting against the window, avoiding the splinters of the ledge, searching for words to use to explain what he had done, the perfect formula that might make him understand and then, therefore, his wife. With one hand holding the curtain away, the other squeezed to the soft sad skin of his pate (it felt — no, impossible), he watched a dignified drunk gracing a doorstep with his arse.

Was it because of Simon that he had run? If so, why now? Was it Charlie? That war was ancient. He worried for his mother, husbandless Dinah, now childless Dinah. He worried for his family and he worried for himself. He worried about the company, the things that Charlie and maybe Leslie might do. Charlie was alone in charge now. Charlie had won. (Why must all roads lead to Charlie?) Howard had given up, surrendered, the family blessing, his father's curse. But that was what he wanted. He had done it freely. He had chosen.

He worried about the past, all the things done wrong and all the things stupidly missed, and he worried about the present with a sick feeling in his belly that he hadn't felt since he was a child. He worried about the future, which stretched out featureless in front of him, terrifying, without horizon. Meanwhile his left hand kept creeping up every few minutes to feel his bald spot. He knew its contours well, he'd traced its ugly appetite for hair each day of his decline, and as his fingers masochistically, automatically homed in, he couldn't quite reject the crazy sensation — it couldn't be, could it? — that it was smaller than it had been the previous day.

— Fuck Charlie!

Who said that? He had, startling himself with his unfamiliar voice. Stronger than the old one, without that whine that had crept in one day like a stray cat and refused to leave. It all came down to habit in the end. He had been the person he was only because Charlie was Charlie and Vivien was Vivien and Simon was in his room and so on.

He would be Howard Astaire now, he liked the dapper phoniness of the name. Howard Astaire returned to the chair. He looked through the photographs, the unlikely, crinkly-edged montage of somebody else's life.

The honeymoon shot of lanky Howard Levy and graceful Vivien entwined in leather jackets with a Cornish harbour scene glinting in faded '60s colour behind them; Vivien in afghans holding an alien-faced baby tightly away from the camera; Simon and Ruth awkwardly dressed as angels for a family event he can no longer remember; Ruthie at fourteen, puberty's hostage bulging through her school uniform; Howard and Charlie, pretending for the camera, standing together in the boardroom like capable executives, like colleagues. And the final one, ancient and brittle, which he added to his small keep pile along with the credit cards and cash and driving licence, Danny and Sam, the laughing long-ago brothers.

It took three matches to get the fire going. The blue flames of photograph ink curled up against the sides of the bin, the business and credit cards flickered and melted and oozed, the loose paper drifted and cracked and jumped tiny sparks into the air.

While he watched the fire build, he discovered more pieces of paper, a first-class postage stamp, a dry-cleaning ticket, someone's unrecognisable name and telephone number on a scrap of diary paper. They went into the fire. Howard wasn't entirely sure why he kept the picture of Danny and Sam, it was probably connected to the same oddly isolated sentimentality that had driven him to steal it in the first place, but that at any rate corresponded to something within himself. This was how it had begun, this photograph, this touch of the past, this was what had made him run.

Then he slept.

4

HOWARD LEVY HAD been something of a dowd. Howard Astaire was a natty dresser. Narrow green suit with thin lapels, black sports shirt, brown suede shoes, a dark blue beret to cover his baldness. Howard Astaire, a dandy, walked slowly, carelessly, swinging his arms and darting his head, his abrupt changes of direction raising angry apologies from the people he bumped into.

Walking was a perilous, thrilling adventure. A shaggy youth with a one-eyed dog made Howard feel as if he was breaking into a thousand pieces. Men with other men made Howard fragile. Women by themselves made Howard bold. Howard Astaire, a sensualist reborn, gulped with love at the women in the street.

Howard Astaire was on his way to attend Howard Levy's final appointment. He would walk the journey from Paddington to Upper Wimpole Street. The day was prettily grey, glittering bicyclists, shiny ribbons of cars, stone buildings glistening with a light covering of rain. He stopped off at a cafe where he consumed a plate of bacon and eggs, considering the whole time without ever deciding on the answer whether he was eating with such gusto because he liked the food or just because it was stuff forbidden him in his most recent life. Then he sauntered along handsome streets, buffeted by people with important things to do, and he tried to work out if the tingling in his scalp meant he felt cold or hot as he sucked into himself all his senses could gather, like a baby feeding on the most mountainous tit or a prisoner just released into an unremembered, shockingly good world.

He held on to the photograph as he waited in the trichologist's reception room. Danny and Sam joyous together at a permanent five to five in the historic afternoon. The photograph was weathered and smooth, smooth as the heads on the people that surrounded Howard. Old ladies with bad nerves and bald heads. Men who had slapped ginger toupees over greying brown sideburns. An adolescent girl shivering, eyebrowless, beneath a bottle-curled wig. Seaside paintings were on the walls, shimmering beachscapes where grizzly little starfish made their plans in the sand, happy sun-blessed places. The photograph was his talisman. It was, even if he didn't like to consider why, important. So was his baldness. That much was indisputable. The rest would have to be added, hair by hair.

— Professor Daniels? Simon Daniels?

Howard Levy had gone by a pseudonym here, for vanity's sake (and if my enemies should laugh at me *here*?). He followed the nurse, her sturdy body upholding its monument of lacquered hair, along the corridor, to stand in front of a closed, numberless oak door. (Another one? Why did life consist of doors that were closed?) He was tempted to burst in, hoping to interrupt Doctor Cohen in an act of congress on the couch with another sturdy white-coated nurse; Howard knocked lightly, waited for permission to enter. The permission came. He entered.

Doctor Cohen's own hair was white and loyally lavish. Doctor Cohen, in his white coat, sitting alone at his mahogany desk, looked through half-moon glasses at Howard's notes, then at Howard. Doctor Cohen sat back and Howard sat down. Howard Levy had suffered from a schoolboyish deference towards Doctor Cohen. The difference between them dated back to Biblical times: put them in robes, surround them with desert, cover their faces with hair, and the Cohens were always priests and the Levites, the next rank down, their helpers. Beneath them both (and much further away from God) were the great golden-ass worshipping masses of Israelites. Let the centuries click forward. Shave them. Besuit them. Cohens are always slicker than Levys. But not slicker than Astaires, Howard was sure that there must be something in the Bible about God's taste for dancing — ballroom perhaps, the tango.

Doctor Cohen was the least unsuccessful of the men who had

taken Howard Levy's money away and given him back no hair. Howard had tried so many things, so many ways. He had given money to charlatans with oily hands and avuncular men with European names who had plundered the alphabet for letters to adorn their business cards with. He had visited hospitals, consulting rooms, basement studios, brightly lit shops in fashionable neighbourhoods. He had answered personal column advertisements and travelled out of London to throw cash at impoverished wizards. He had changed his diet. He had eaten only fruit and vegetables, he had eaten only proteins, he had eaten everything but dairy products, and then he had eaten only dairy products. He had submitted to scalp massages with aromatic oils. During a meeting with five evangelical ex-financial brokers, he had prayed weepingly to God and bought bucket-loads of proprietary ointments containing the necessary hormones. He had travelled twice a week to a private hospital in Wimbledon to be abused by a Stress Therapist. He had received a grubby hypnotherapist in his own house, regressed briefly to a past, hairy life, and his house had been burgled three days later. He had grown his hair long and pulled it tight over the abyss and tied it at the back with elastic to make a ponytail. He had affected wide-brimmed hats. A theatrical wigmaker had measured and fitted him but Howard had never returned to receive his 117B-auburn. For two weeks he had sprayed his hair hole with an aerosol but then abandoned that when he had endured a business dinner with a face smeared with dye that had run in the rain. He had gone three times a week to a basement in Chelsea to be strapped into iron boots and hung upside down from an iron bar. None of it worked. No dice. No hair.

Each morning, the bow of the head, the grim look into the mirror, the naked cancer growing, spreading out from what once had been a coin-sized dot in the centre. Horror as the clump at the front became separated from the rest, an island connected by two narrowing bridges to the mainland at the sides.

— Your hair looks good, Professor. Please. Sit on the couch. Let me examine. We prescribed you a keratolytic, didn't we? Are you still on it?

— I'm afraid. No. I'm not.

Howard sat on the couch and, as Doctor Cohen rummaged through his hair, he had to battle down the return of the deference. It was a situation that might require rudeness.

— Professor. I had a message for you. Or maybe you had a message for me . . .?

The trichologist always had been prone to lapses of memory. But the precious connections from island to mainland had been maintained under his care so he had earned the right to be senile.

— There's a new drug on the market, you might want to give it a whirl. How's the family? Taking them somewhere special this year?

For a moment Howard was tempted to invent a family, make it happy, and send it off somewhere special.

— How's the Institute? Excuse me. I forget your avenue of expertise.

The Institute. Think of all the studies that the mythical Professor Daniels might excel at. Howard couldn't remember what lies he had given in the past and anyway Doctor Cohen did not require any answers. The amiable question was just a tool of his trade, like swab and fingertips and ointments.

— I came, actually, about my hair.

— Of course. The condition has stabilised, hasn't it?

— Stabilised? No. It hasn't.

— Message. I have it. Your cousin, I believe. A Mister Charlie something. He asked for you to get in touch should you have the time . . .

Horror tapped at Howard. The uselessness of pseudonyms. Perhaps even the impossibility of change. As long as all roads led to Charlie he was trapped. Cut away this doctor, cut away Charlie, cut adrift.

— He says you will know where to find him. Very good. I am going to write you this prescription triple strength. Your scalp is showing signs of life. I am to move shortly. Perhaps even disband the practice. The rent is becoming a problem. Punitive.

— Oh?

A warm feeling soothed Howard. This was good. Other people's troubles. He put his beret back on.

— The building has been sold actually. A new landlord, fast-talking boy with a telephone and a sports car and plans. I told him about my business. I even tried to interest him in some treatment. I treat some very eminent people here as you know. But oh no. What's good enough for admirals and cabinet ministers and music stars and yourself of course, professors, isn't enough for this pipsqueak. He'll raise the rent, all the tenants will be forced out, that's the plan. Then a new lick of paint, maybe central heating improved, new tenants, higher rent. The building won't be filled. I guarantee it. The fellow's a fool. And bald too, as a coot. (He pronounced the word with care and only slight derision, sending it off into the world with a smack of his lips at the end.) He'll let out the space at reduced rates, the bank will get after him, he'll go bust in the end and the building will stand empty with everyone out of business. My brother, you know, died in Spain fighting Franco.

He said it as if it was a recent event, head slightly bowed, dewy eyes searching for fraternal comfort in the bare opposite wall.

— I didn't know. I'm sorry.

Doctor Cohen, despite his spiritual rank, was going crazy. The death of his brother sixty years before, the loss of his practice, or maybe just the shadows of age. Howard quietly rose from his chair.

— Everything should know how to tick over. What's wrong, may I ask you? with *nice*? A working man has a right to his living, I get mine, you get yours, we all have time left over to examine the world, discover a little happiness maybe, share our joys. But no. Crazy greed. The big haul. Fuck 'em. I'm sorry. I am developing a tendency to ramble. And all the time you sit there patient patient waiting for me to talk about your hair. Fuck you too. Excuse me, I was making a wisecrack. You might be interested in calcipotriol perhaps? It's very in vogue these days. Maximum dose sixty millilitres a week. We don't want any hypercalcaemia do we? Professor Daniels? Did I tell you your cousin was looking for you . . .?

But Howard was gone. Swinging oak door. Footprints in lush

corridor carpet. The front door pulled open and crashed shut. Litter scattered. Puddles disturbed. A taxi swerved, reached the Bloomsbury curve, window down, taxi driver's face (peaked cap above, swinging medallion below), *Where to, guv?* Taxi door shut, cab swerved back into traffic. Like his son, the world was getting to know Howard by his absence.

HIS NAME IS, was, Simon. The boy in the bedroom. An earnest child, too sensitive for his own good, twelve years old. Go back. To when the signs first showed. He was nine, a domestic scene. The family at home. The family united.

They were eating dinner around the kitchen table. Faces down, intent on food, that safe thing. A light breeze through the open window. Sauce stains on work surfaces, magnetic letters on the refrigerator door. The pinboard beside the wall-mounted phone, cab companies, doctor's appointments, cinema timetables, important school dates, postcards from hot places pinned promisingly up.

— Why is he called Porky?

— Excuse me, darling? Why is who called Porky?

— It's a joke.

Simon's solemn face, round eyes expecting disappointment, unspoilt skin, heart-achingly thin neck (when he turned his head to reach for juice, he showed the odd twist that his hair tapers down into, miniaturised image of Howard's own), dollop of tomato sauce trickling down the cleft of his chin, narrow in his sports shirt, thin body that hadn't worked out yet how to protect itself from harm.

— Be quiet, stupid! hissed Ruthie.

She was feeling glamorous in the superiority of age — and sexual know-how too, we suspect. Not quite a beauty, features too coarse (genetic reminder of a Polish border town), body that time would prove to be dumpy, but she carried herself well, with style, and there was something to her that was entirely separate from either of her parents. She was teaching herself to become her own image of a great lady, eating carefully, elbows tucked neatly to her sides, hands that held her knife and fork moving as if they were playing some light airy thing on the piano. Simon hated Ruthie and Ruthie hated Simon and that was probably as it was meant to be.

Simon lowered his head in his expected suffering and Vivien tried to help him out.

— Tell me, darling. Why *is* he called Porky?

Ruthie yawned extravagantly, looked at Howard as if they were at some nineteenth-century Russian ball and they alone shared a superior romantic complicity. Vivien leaned forward, encouraging her son. When had she acquired that sour thin edge to her voice? Her face was lined with trouble even then and that was Howard's achievement.

Simon looked at his mother, suddenly struck by panic. He held his cutlery in tight fists, he mouthed quick inaudible words as if to convince his family he was practising his punchline when it was clear to see that he realised he had muddled the telling and got the joke wrong. His face pinked, his mouth pulled into a painful smile, he suddenly shrieked out an impression of a police siren as if that was a dangerous, exciting enough sound to get him through this. He checked the expressions around the table and his face went from pink to red. Ruthie suavely knocked away one of his elbows from the table and a fork went spinning to the tiles and the arm flapped down into his lap. He could have taken this opportunity to lose the subject. He could have attacked his sister with his little thumb-tucked fists and in the ensuing battle nobody would have returned to the joke. But he was too stupid or maybe just too scrupulous for that.

— I don't remember, he finally whispered.

And then he covered his shame and his guilt at not having provided his family with something funny by abruptly squelching his face down into his warm pizza.

They gathered, in safety, in the TV room. A quick skirmish, not heartfelt, just a matter of pride and tradition, over which channel to watch. Howard's was the casting vote, and, to madman mutters from Simon (splatter movie) and grande-dame groans from Ruthie (pop music programme), Vivien's mild choice became the winner. Hollywood black-and-white, scared innocent faces lit from below, men in hats talking fast on drug-store telephones, women wrapped in ballroom silk drifting down party staircases. Howard sat with Vivien on the sofa, husband and wife together, not touching; daughter on the floor practising some new weight-reducing exercises; son curled miserably in an armchair, eyes flicking to the ceiling in disgust at the alien world of 1940s Los Angeles, controlled musical belches popping from his mouth, as he planned, no doubt, some future dark revenge against his sister.

Vivien's eyes brimmed with tears, the easy sentimentality of Hollywood. The tears dripped eccentrically down, guided into mazy patterns by the cracks on her cheeks. Howard stared at the control knobs on the TV set and tried to remember how his life had worked out this way. Ruthie looked over and then quickly away, appalled by the shameless display of her mother's softness. Simon closed his eyes. Things hurt him. He got up. He went to his room. He remained there for a day.

The signs were all there, if anyone had been looking for them. And they would have been there earlier, a week before, a month, a year, a lifetime. Perhaps it was all genetic anyway — the need to run, a passion to hide, you could trace it in the bloodline, look at what went before, look at dead Danny. So maybe it was genetic, a thing in the blood that bubbles up through the head to turn them into hermits, or maybe it's not.

It happened gradually. Simon was twelve, a stubborn boy, with the fluffy beginnings of a beard, who was going through what was, they all decided, an awkward phase. He was combative in conversation, sour in repose. He showed no interest in school. He had been dropped from the football team and he didn't seem to care. Fertile acne thrived upon his fluffy face.

He didn't go out as much as he used to. At meal times he would pick at his food and stare into space and if anyone spoke to him he would either answer with a simple *No* or act as if he hadn't heard a word. He started to take some of his meals up in his room, and Vivien would save him the effort of coming downstairs for the tray. Instead she'd bring it up even if he hadn't asked for it, hoping that maybe in the moment of passing the tray through she would find out something of what was troubling him (or at least he might remember that he used to enjoy talking to her, that, in quiet moments, years before, he would brush her hair in the bedroom, they'd count them out together, the hundred strokes, and she would whisper, hoping her husband wouldn't hear, but he usually did, that she wished Simon could be like Peter Pan and never grow up). Sometimes mother and son would talk, but only briefly, and of nothing consequential, but often he would just call out to her to leave the tray outside on the landing, and sometimes she would just leave it there without him needing to tell her to.

The final argument was over this: he had wanted a bicycle for his birthday. A ten-speed racing bike, drop handlebars, light-frame aluminium, he'd seen the model he was after, fallen in love with it, steel-grey, the colour of a killer's eyes, with a green flash down its central stem. He'd discovered it in a bicycle shop in the West End, made pilgrimages to it on Saturdays, watched it come down in price over a space of three months, even sat on the adjustable saddle a small number of times before the assistant got wise to him. He dropped hints for birthday and Christmas, hints which had fallen where he dropped them and remained there and rotted; and instead he cried tears over a leather football (coloured orange, not even white like it was supposed to be) and an electric razor in a black plastic carrying-case.

The night he took to his room, he finally came out with it, hints forgotten, true desire spoken.

— I want, he said, a bicycle.

His mouth twitched with frustration and need, his eyes darted around the kitchen table, looking, hoping, for an ally.

— You have a bicycle, his mother said.

Simon shook his head, fought back the tears that were threatening his eyes. Yes, he had a bicycle of a sort, but in the world of the adolescent (why hadn't they realised how ruthless a world that is?) it was a laughable sort of bicycle. Three gears only, upright handlebars, not even a mountain bike's handlebars, and it was too small for him, he tried to explain. Howard had suggested raising the saddle. Simon shook his head, a couple of tears started on their sad journey.

— People laugh at me. It's not the right bicycle. It's too small for me. It's old fashioned. I — I'm not a *postman*.

His sister intervened. Prissily she said something about how people can't always get what they want. Pompously she said something about the things that she wanted but had to wait for.

And Howard could only sit there, looking at each of these strangers and trying to give the impression that it was a pleasure he would strive to be worthy of to be surrounded by such amiable performers.

— I don't want, I need! I NEED a bicycle, this bicycle, NEED it, it's cheap, and *it's the last one in the shop*.

— Get a job, his sister said.

Vivien looked at Howard. Howard struggled to recognise her.

— Next year, he finally had said. It had seemed like the right kind of thing to say.

— *But I*—

— Let's not go on about it. Let's not be boring.

Vivien offered Simon what she hoped was a sympathetic wince. Ruthie smiled at him in odious triumph. Howard wiped his mouth with his napkin and gazed at the kitchen wall.

— I'm going to my room.

— Wait until you've been excused, darling. You—

— *I'm going to my room*.

Simon rushed from the table. He must have been desperate to get away before he showed his sister any more of his tears. He dropped his napkin to the floor, and the side of his leg knocked the spoon out of his soup bowl and before it had landed he was already on the stairs.

They'd heard him crying himself to sleep that night. He'd always been a sensitive boy, the sight of a beggar in the street would empty his pocket of money: pain hurt him, it brought tears to his eyes; and it could be his own pain — a knee skinned on a breakfast-tray toboggan run on the Heath in winter — or it could, with an equal effect, be somebody else's: an alcoholic clutching a leaking bottle; a dog limping around a corner; even the sight of a shopping trolley abandoned in the courtyard of a block of flats, upside-down, its wheels spinning from the casual touch of a passing boy, could make his throat raw.

He stayed in his room throughout the next day as well. He listened to phone-in programmes on his transistor radio (LPC-manufactured, circa 1968) and he ate food from the trays Vivien left outside his door. He didn't come out of his room again.

Simon had been in his room for nearly a year when Howard ran away. The family had tried everything they could think of. They tried to wheedle him out. They tried to punish him out. They tried to starve him out.

The situation didn't change, only developed its own rules. He ate from trays left outside his bedroom door, careful only to retrieve them and deposit them when he was sure no one else would be on the landing. A couple of times Howard caught a shadow glimpse of his son's peach-fuzzed face peeping around the corner of his implacable door, then quickly retreating when he sensed his father's presence. Simon was never seen to use the bathroom or the toilet, and he always covered up his traces. His school sent a teacher to call, and then an inspector from the local authority, and, with an endowment there, a letter here, they managed to convince the authority that the boy was being adequately educated at home and there would be no need for any official action.

The night he took to his room they decided sweet-toothed hunger would bring him out. So they ate their pudding in the dining room down below and loudly made noises over how delicious it all was, while he lay on his bed, probably imagining the journeys he could take on his missing bicycle. He didn't come down for pudding and when his mother passed by his bedroom in the morning, the tray she had left out for him (containing a glass of milk, a bar of chocolate, and two profiteroles half-buried in cream on a plastic Alpine scene) had not been touched.

Later that day they tried his door. It was locked. They called out his name, his full name, not the customary endearments they had learned would annoy him, and he made no response. His friend from school came to call and was told to go away. His sister knocked on the door and wasn't even acknowledged.

The next day they tried to starve him out. Vivien cried in the kitchen as Ruthie sternly removed the sandwich and the glass of milk from the tray and replaced them with a glass of water. He wasn't given breakfast or lunch or dinner the next day. He didn't come out.

They tried to bribe him out. Howard talked briskly of bicycles and computer games and horseback riding and grands prix. Vivien prepared his favourite meal (pepperoni and onion pizza followed by chocolate marshmallow ice cream) and left it for him on the landing, with a note folded up and propped beside his glass of milk. Ruthie slipped miniature Swiss chocolates beneath his door and promised to do the washing-up for the next three weeks. Howard, the impotent, foolish dictator, ordered him to speak. He didn't speak. They sang — his father, mother and sister in a line on the landing, accidentally in order of height — Simon's favourite songs, Sloop John B and Molly Malone and Life's A Gas. They sang the songs and there was no call for an encore and they went back downstairs again.

They thought his grandmother might lull him out. Howard had always been a little jealous of the love between Dinah and Simon. It excluded him. When Simon was small and his grandmother came to visit, he always ran toddling out of the house towards her, his arms in front of him, delight on his face, no reserve. She stayed with them sometimes in those days and there was always that world-ending horror on Simon's face when he realised she was getting ready to leave. It was as if something had entered him, in the shape of her fingers, filling a special place inside, and when it was removed nothing else was the right size to replace it, so it just remained there, empty, getting colder. Dinah sat, gracefully curled, on the landing out-side his door and told him gentle stories of the family past and waited for him to come out. He did not come out.

It was peaceful there, outside his door. Vivien fell into the habit of sitting there to talk to her son, to tell him about things that had happened in the house and sometimes in her heart too. He did not come out.

They tried electricity. One weekend, Howard went to the fuse box and turned the switch and the house was in darkness apart from the rooms they had prepared with candles. They ate food brought in by motorcycle riders from outside, pizzas and ham-burgers and wonton soup and deep-fried chicken. Howard and

Vivien and Ruthie played parlour games in the flickering candle-lit rooms. The absence of electricity did not drive Simon out.

They bought him his own television set. He silently rejected it. Sometimes, notes would appear, in Simon's rough handwriting, requesting things. He requested a store of batteries for his transistor radio. He requested certain books. He requested a change of clothing. Later, after the incident with Ruth, he requested a chemical toilet so he would no longer have to make the dangerous walk from his bedroom to the bathroom at night.

They tried model airplane kits, Messerschmitts and Spitfires and Hawker Harriers and B-52s. There had been a time when Simon had lived for these, long nights in the kitchen solemnly putting them sloppily together, ridges of glue rising out from between the polystyrene panels, decals stuck on a little askew, pinched in the middle, not quite flat. When the planes were complete and painted Howard would suspend them by string from Simon's bedroom ceiling. Together they sometimes composed air battles and dogfights. Simon in his seclusion did not want model airplane kits.

Simon wanted books. He wrote their names on scraps of paper, pushed them out into the world from beneath his door. And Simon wanted pencils, with soft heavy points. And Simon wanted batteries for his radio. And Simon wanted a tennis ball, but they never found out what he wanted it for, they never heard it bounce or roll. And he wanted a computer too, with games to play on it and a modem to connect him, bodyless, to the world. They did not give Simon a computer.

They thought a lack of voices would drive him out. They thought loneliness would send him back to them. They invented an earlier time and they called it happy family. They thought a dead radio would make happy family.

So they tried batteries. No, they said, you shall not have four AA batteries, HP7, one and a half volts, and neither shall you have an extension cord to plug into the wall socket. When your radio dies, it shall remain dead.

They tried books. No, they said to his frequent requests. We will not provide you with any more books. Not medieval war books, nor private eye fiction, nor literary novels translated from the French which contain ecstatic descriptions of the labia,

67

nor superhero comic books, nor the lives of the saints, nor books about the history of engineering or British football in the 1950s.

They tried silence. Vivien stopped dallying outside his door, telling him things. Simon stayed in his room, invisible. They resumed their supply of batteries and books and talk.

They found him in the bathroom one night. He had been peacefully alone for nearly three months and contentment must have made him careless. He was lying in the bath, surrounded by bubbles, listening to a radio short story sent in by a listener which was about a dying angel called Fred. He had neglected to turn the key in the door.

His sister was the first to discover him. She had been lying awake in her bed imagining herself to be a princess in Russia before the revolution. A handsome captain of hussars with a glorious moustache and duellist's eyes was bowing before her and inviting her to dance a quadrille. The radio next door approached its loud climax and tore apart her preferred Cossack world. She was about to shout when she realised what the bathroom radio sound signified. Very softly, Ruth climbed out of bed, wrapped herself in a bath robe, and tiptoed to her parents' room. Ruth had opened the door and lifted a finger to her mouth and jerked her head.

Ruth and her parents went softly down the landing. They stood in front of the bathroom door. For a while they listened along with Simon to the conclusion of the story about the dying angel named Fred. Simon was ready to get out of the bath and this is what saved him. He reached for the towel he had ready on the bidet beside the bath. He stood up in the tub and pulled out the plug for the water to drain. He lifted his right foot on to the rim of the tub and set about rubbing his skin dry so he would leave no watery traces behind. Vivien reached for Howard's hand. They tracked their trapped son by the meticulous sounds he made.

It's not locked, said Ruth. She and her mother both kneeled by the door, trying to look through the keyhole (which Simon had wadded with toilet paper). They started to rise. Howard turned the handle. Ruthie stood up. Her hands were poised to enter her brother's private world.

Simon was nearly finished rubbing himself dry when he must have seen the door handle start to turn. It turned very slowly, then more quickly, and a stream of light over the tiled floor became, suddenly, a lake. He jumped to the door — splashing water all around, knocking the radio into the bath, he slipped on the wet tiles, half ran, half slid to the door which his shoulder rammed against, trapping two of Ruth's fingers, a girl's scream, a mother's gasp, he pushed the door harder, not knowing what the impediment was, the scream became higher-pitched, his mother called out his name, he must have seen the fingertips, red, swollen, bulging with blood; he opened the door a fraction, the fingers retreated like half-dead animals looking for shelter, then he slammed the door shut and they heard the lock turn.

Ruth started to cry in relief and pain. Her mother held on to her hands and kissed the blood-filled tips. Howard had wished for a thought of what to do. Ruth knew what to do. She screamed abuse at her brother. *You've broken my fingers, you bastard*! she shouted. *They'll have to be amputated you fucking cunt*!

Vivien and Howard took Ruth to hospital, the casualty ward, where, after they spent a solemn frightened hour surrounded by the dead and the dying and the broken and the merely scared, a doctor nodded at her injury and quickly burned a round hole through the centre of each of the damaged fingernails, and Howard had looked away because he couldn't bear to see his daughter's exposed nerves blackened with blood.

By the time they returned home, Simon was back in his bedroom. A virginal spot of Ruthie's blood on the tiles was the only trace left behind. The bathroom floor had been dried. The bathtub had been cleaned. The drowned radio had been retrieved and now defiantly played, louder than usual, behind Simon's locked door.

Shortly before Howard ran away he bought his son a bicycle. Vivien tied a big blue bow around the handlebars. Howard left it propped against the landing wall outside the bedroom, described it to his son through the locked door: lightweight aluminium frame, drop handlebars, ten racing speeds, grey with a green flash down its central stem; but it remained there, ignored, untouched, because Simon had gone beyond bicycles.

6

HOW TO DO IT? A sudden lurch? Jab his arm painlessly against hers, make sure the lager spilled, but not too much: *Oh excuse me, I'm* — then brush her wet sleeve, give her a smile, connect, *I'm so sorry, I'll* . . . And then they would fall into conversation, avoiding the damp beer patch on the seat between them as they found their common ground, an old TV programme they both enjoyed or their shared faith in the goodness of dogs.

It had been so long. He couldn't remember the last time, it could even have been Vivien in Tottenham Court Road where the Pink Floyd used to play. Ever since then he had been carefully, comfortably, almost completely faithful. Should he turn to her, feign an astonishment that they should meet like this, so chancefully, offer up somebody else's name? *But you're not? Why that's uncanny. I could have — even your hair is the same.* And then, breaking off the discussion of hairstyles and fate, he would buy her a drink; and when he returned from the bar the conversation would inch out of the philosophical realm and into the physical.

Howard turned to the woman sitting beside him. She was stocky and she wore blue dungarees with an inside-out yellow T-shirt beneath. She had short blonde hair with a red shine to it and she had come in alone but spent the past ten minutes with two moustachioed Irishmen who had left her now for the pool table. Howard said hello. She said hello back. He smiled at her and she at him. He shifted a little towards her, there was no damp patch to separate them.

— Would you like a drink?
— Are you married?
— Yes.

— Where's your wife?

— I don't know. At the house? I don't know. Would you like a drink?

— A pint of lager and Guinness for my friends and what about money for the pool table?

Her name was Carol. Carol had a voice that had been ruined by cigarettes, and she had been pretty once and she carried herself as if she still was so he believed her. They played games of pool, doubles; he partnered her against the Irishmen, drooping moustaches, testosterone brothers, who both acted as if she belonged to them once and would again if they could only make their voices loud enough and their words ugly and abusive enough. And sometimes, when she was about to take a shot, one of them would stroke Carol on her denimed arse, and a lot of the time all of them made fun of Howard, his accent and his beret (*Your shot now, I believe, Lord Pee-air*). And at the end of the evening somehow the Irishmen were gone and he and Carol were outside the pub and it was raining and they each had their feet wide apart to keep from falling and their foreheads were rocking together and both of them were laughing.

— Where do you live?

— Nowhere really. A hotel. It's round the corner. I could take you there if you liked. Where do you live?

— Local. On the top floor. I'd remember if I saw it.

— Would you like to come back to my hotel?

— You're no fucking gentleman. You going to take advantage of — of — of — of my *condition*?

— Only if you take advantage of mine.

She liked that reply and so did he even though it surprised him to hear his voice say it. She went back with him to his hotel.

They walked quietly up rickety stairs, along a corridor decorated with peeling tourist board posters (all for places beginning with the letter A — Austria, Australia, Angola, Antarctica, Arbroathshire, Andalusia, Angora, Athens, Amsterdam, and a final one for antique Abyssinia), past bedroom doors that he had never seen open, and into Howard's room at the end of the hall.

— It smells in here, she said. Have you been cooking?

— No. Yes. Sort of, he said.

They lay on the bed and his face was full of the creamed corn scent of her as they fought a war over her dungarees.

— No, she said.

He said nothing. He stuck to his task, pulled apart some poppers, bruised his hand on rivets as she battled to reclasp the poppers he'd popped, and he felt like he was dismantling a piece of machinery.

— I thought you were gallant. Please. Brother Howard. I've just had an operation. I don't want you to see.

— What kind of operation? he said.

— Please, she said.

Carol was strong. Carol had small hard hands with sharp tearing nails and powerful legs and jaws, and her muscles had been trained in combat while his had grown soft in a clean office and a remote home. He had to surrender his first objective, of getting them both naked at the same time, concentrate instead on attacking the weak defence of her dungaree poppers, fight to subdue her with the weight of his body, pin her legs down with his, accept the damage done by her hands and her teeth.

He was not gallant. He was relentless. He won his war, with the acceptable losses of a few blood-trickling skin wounds to his shoulders and ears, and now she was naked beneath him apart from the sports socks on her kicking feet and the sky-blue knickers tugged tactically down to just below her knees and the inside-out yellow T-shirt pushed up above her navel; and she was shaking with fury or it might even have been modesty, and he was a thousand feet tall from the joy of doing something he felt sure about. He clutched her head between his hands. He tried to conciliate her with kisses.

And finally her face stopped struggling away from his and through her skin she let him stroke the curves of her body, and he felt as if he was touching his own flesh. No matter how hard or soft he stroked, even with his eyes open to remind himself of the difference between them, he was stroking his own anxious face, his own high cheekbones and dimpled chin and far-apart lips and icy edges of teeth, his own blonde hair with a touch of red to it, his own smooth hard panic-muscled shoulders, his own wide breasts, the sudden inward curve of his own waist, the soft sides of his own arse sandwiched between a shallow mattress and a man's weight, his own scarred belly, his own cunt, gently pulsing, gingered with hair. And as he removed the rest of their clothes — breaking off to follow the lines of biceps and shadows on throat and the downy skin of once-tanned thighs — she shivered and said something that he couldn't hear but which sounded glad as she reached for a condom from a pocket of her dungarees.

He refused to wear it and she was too tired or drunk or proud to make him, and as he fucked her he thought almost entirely of her and seldom of his wife, and after he ejaculated — half of it inside her, the rest of it, after she heaved him out, pulsing onto her belly and rolling into the long dry purple riverbed of a recently carved scar — he fell quickly asleep.

7

SHE WASN'T THERE when he woke. He rubbed his head, pulled the covers tighter around his body, which continued to shiver, rubbed his head again and then his face. His mouth tasted like an ashtray washed with vinegar. His teeth were blades of pain. The harmony of stomach and bowels and sphincters had never before been so clearly revealed in all its true complexity and precariousness. A £10 bank note was propped up beside him on the pillow, with a message written in mascara over the head of the Queen — *Lent £5. Luv C.* He turned away from the crispy patch on the bottom sheet. He felt very bad. Had that been what he truly was last night? Finally undisguised, a flesh monster, liberated by drink and a new beret? The sight of a small crop of Carol's yellow-red hair on the pillow beside him filled him with sadness. On the bedside table was the photograph. Howard briefly stroked the unruined face of his father before he rubbed his own head again, was that *stubble* pricking up beneath the surface?

He would have to see Katz. Do it in person, otherwise the lawyer's legendary laziness might subvert the plan. It felt like cheating but it was his only choice — money for everybody, then he could forget all about the accidents of the past. He couldn't go home, that was not possible now, if ever. Not until he found out who else was inside him along with the flesh monster; he could hear them inside growling, felt them, saw them, ranks of sun-starved convicts itching with desire, grinning as the chains of a lifetime rattled, loosened, creaked.

Howard climbed into his clothes, not so new any more, not so dapper. He called the company from the payphone in the lobby. He hardly had to disguise his voice, his hangover did the job for him, but he held a handkerchief up to his mouth as he barked his way to a connection with Katz. Katz was obliging and ironic as ever. The lack of surprise in the lawyer's voice intimated either that this call was not entirely unexpected or that there was no room in Katz's world for surprise. Howard told him what he wanted. Katz suggested that they might meet at a colleague's chambers near Gray's Inn. Then they negotiated over the time. Howard wanted to meet immediately. Katz suggested that next week might be more convenient, given the amount of paperwork, not to mention telephone calls. Howard offered one hour. Katz reluctantly accepted two.

Howard left the photograph in his room. It was a hostage, guaranteeing his safe return. He gave it a view of the street outside, where men sold broken things out of supermarket trolleys and customers went into the bookmaker's and the timber yard but seldom, Howard was sad to notice, into the Hair Academy (*Lace front hand-made human hair £299, fitting included — Our Hair Piece Consultant has 65 years experience*). In the upper storeys of the Hair Academy shadows slipped behind the roughly boarded windows.

It was raining again. Heavy cold rain pouring down, washing through everything, purifying. People in shop doorways with their collars turned up inspected the weather as if it was everything they had ever grimly expected from life. A few pathfinders muscled along the street behind umbrellas held like jousting lances. Howard, unprotected, baptised, strolled, and he turned a corner and the connection was lost. The photograph in an empty room waited for him, or another Levy, to come back within its ambit.

He returned with the money and the rewards of a box of chocolates and a transistor radio. He had gone out into the dangerous world and escaped from it again with a stack of money besides. On Katz's colleague's empty desk there had been the asked-for wad of notes, several documents for Howard to sign and, monstrously, a note from Charlie who still knew how to predict his movements. Howard had signed the documents, thereby cashing in, at a penalty, two pension plans and three life insurance policies. Katz had suggested he could sell his shares in the company too but Howard was not quite ready to do that even if the prospect appeared delightfully irrevocable, end of story. Katz pushed the note from Charlie towards Howard. *He wants to see you*, said Katz. *He thinks he has things to discuss that might interest you.* Howard had counted the first installment of his cash and he had asked again to be reassured that his family would receive their money on the first of every month; and he had left the note from Charlie where it was, untouched, this emanation of the enemy, on a stranger's desk garnished with a few stray flicks of cigarette ash and several strands of short dark hair.

The rain had stopped, of course it had, the sun shone upon him and his unwritten fate. The rest of the day was a tender passage of rich truffle chocolates and gentle radio music. The floral curtains were drawn. An afternoon street light flickered roguishly outside. Inside, cellos and violins surrounded him with the sad soundtrack of heroism. He ate the chocolates slowly, breaking the thin dark shells with his upper front teeth, dipping his tongue into the bitter, slightly metallic centres. He sucked and he chewed and he smeared his chin and fingers with chocolate stains. He lay naked on the bed. The webwork of hair on his chest and belly and the lattice of raised blue veins on his legs and feet and the map of cracks in the ceiling above were the indecipherable prophetic pictures of his own glorious future.

8

HOWARD MET CAROL again at the pub. She had been sitting alone in a short white dress, waiting for somebody, and as Howard got quickly drunk, together they pretended she had been waiting for him. She had done something strange to her hair to make it spiky. Her legs were crossed, one white stiletto shoe tapped against the stem of the table, her thighs plumply squeezed one another. When she smoked, she painted the filter tip red with her lips each time she plucked the cigarette away to blow brief, perfect smoke rings.

— You keep moving about, she said. Have you got fleas?

Clamouring inside him were a thousand possible selves. Howard felt ready to explode from the pressure of their seminal battle to be the one to emerge and get laid first.

— Look, there's Mickey. You better make yourself scarce.

One of the Irishmen from the night before had come into the pub. He was wearing jeans and a country and western shirt and his hair had been wet-combed back over his bald spot.

— He hasn't seen you yet, said Howard. Let's go.

This at least was clear. He didn't know if it was the alcohol or the first glimmering of free will but she was not to be with Mickey. She was to be with Howard.

— He's who I'm meeting here. Sorry.

Howard took Carol by the arm and as he steered her out of the pub Mickey never saw them go or the eyebrow she raised or the red marks Howard's fingers impressed on her arm.

It was cold. Howard and Carol walked to the concourse at Paddington station, where they picnicked on baguettes and cappucinos, and then they wandered along a parade of shops

— closed-down cafes, a camera place, a tapas bar, a massage parlour called Scandinavia.

— It's been fun, said Carol, but I have to go soon. I'll be late for Gavin. His auntie's babysitting and she's got enough on her plate, it wouldn't be fair. And Mickey. He'll be out for you you know, Brother Howard.

— Why do you call me that?

She burst out laughing as she considered something that was too funny to share.

— What? What is it?

— Your *hair*. It's long all around the sides and the front and the back and really short at the top. You used to shave it at the top, didn't you, like monks? It's called something.

— A tonsure.

— I don't know. How long were you a monk? And why did you stop being one?

— Why? I, you, miss things.

Howard liked the idea of being a renegade monk. A contemplative man, holy even, who had penetrated the mysteries of God and had chosen the flesh instead.

They came to a bench. They sat, extwined, upon it. Together they admired the view of mini-cab companies and prostitutes and hotel families sitting on steps. Dodgy men standing in groups occasionally took turns to kick at the crushed beer cans scattered along the pavement.

— Stuck away like that I'm sure it has advantages. What's it like? Being back in the world.

— I — how —?

— I would imagine, how would it be . . .? (She shut her eyes the better to imagine.) It must be exciting and it must be scary both. And horny too, I reckon.

— It is.

— I always wanted to fuck a monk.

Howard lifted his hands to his head. He rested the balls of his hands against his temples, his palms against his skull; he slowly lowered his fingers to touch the bald patch at the centre. And — no smooth skin, no shaming nakedness — instead of the bald hair hole, there was a full growth of short stubble, agreeably sharp against his fingertips. He wanted to yell. He wanted to holler. He wanted to scream out blessings but he didn't know who to praise. Except, maybe, the god of irresponsibility.

Howard got up from the bench. He found a space between the beer cans and knelt on the pavement with his chest against Carol's knees. Howard leaned his head forward between Carol's legs and the men across the road cheered.

— Are you going to be sick?

— I, could you do me a favour? Tell me what colour it is?

— What do you mean? Oh I see. Not positively, the light's not very good, but I'd go for brown, *maybe* black, maybe, but probably brown.

— No grey? (his muffled voiced plea).

— No grey.

Howard stood. The men across the road booed. Joy was in his heart. He could feel on his face the beatific expression of a monk.

Howard bullied Carol to her feet. He hailed a cab. He told the driver the address off Western Avenue and pushed Carol in ahead of him.

— Where are we going?

— The scene of the crime.

— I don't want to be late —

— For Gavin. Yes I know.

— Are you making fun of me? Because if you're making fun of me . . .

— I'm not making fun of you. Look. This picture here. You see this? The little guy is my father. His name was Daniel. The strong-looking man was called Sam. He was Daniel's brother. Charlie's his son.

— Who's Charlie?

She took the photograph from him. Danny and Sam, the founders of the family fortunes, smiling and dancing their Unbreakable Comb dance. She looked about in the wrong places for a light switch and he found it for her and she compared his face to his father's.

— Your eyes are the same.

— I look more like my mother.

— And the shape of your face. The chin is identical.

— Yes.

It made him happy, more than happy, it exhilarated him to learn that his chin was identical to his father's. And then it made him feel jittery, like something terrible was about to happen.

— He had a weakness for women with red hair.

— My hair is red. Sort of.

— I know.

— Oh. Is that — ? Tell me about him.

— He died the year I was born. In a fire.

— But you must have been told stories.

There were a million stories with Danny as their leading man and most of them contradicted each other. Everyone had stories to tell about Danny. Dinah told the best ones. Freda told the sternest ones and Charlie the dirtiest ones.

— I never met him. I would like to forget him.

They were approaching it. There the narrow semi-detacheds, painted skins grimy from a century of cars. There the block of flats — smashed windows, temporary wooden doors — on the long banking corner of the road. Tube stop, petrol station, Happy Eater, and just ahead, stupidly squat, hugging its patch of earth, never daring to reach up to the sky, the factory building.

Howard walked with sleepy Carol down the lane, through the factory car park, past the main business entrance, past the old sign that claimed THIS SITE IS PATROLLED DAY AND NIGHT BY DOGS!, up the concrete staircase to the workers' gun-metal door, where they stopped beneath the twenty-four-hour halogen lamp. He pulled out the photograph from his trouser pocket. Sam's hair bristly like Charlie's, light grey in the monochrome photo, ginger in life. Danny's hair, black, curly, tumbling over his forehead into his sad brilliant eyes, an unruly testament to genius.

Behind the building, a brown stony field, a clump of trees, a dark stream. Over there, in the wilderness, near the railway lines, was where the runaway convict had been found by men with dogs. And all of it a playground for Howard and Charlie (he was still called Lawrence in those days), warring cousins in school holiday shorts. Trees to climb and hope the other falls. Field to risk bicycle tyres on, the Tour de France, wheeling around and around; and before that, medieval jousting arena.

Start at opposite ends, each with a metal dustbin lid on one arm and bobble hats for helmets. The signal given by the imaginary fair lady, an imaginary silk scarf in her own exquisite colours drifting to the earth — in fact a rocket ship countdown yelled out by both of them — and then, push off, gather momentum with scuffing feet, pedalling fast, the enemy directly ahead, mixture of terror and triumph in their hearts, reach the middle of the arena, a clash of dustbin lids, bicycles smash, crash, buckle, both clatter to the ground, pebbles rip their knees, bite into their faces, ruined wheels spinning, and each of them, lying so close to one another on the earth, refusing to cry until he's heard the other give way first. And then Freda's voice, harsh from her office window, calling them to grim punishment.

And then there was the time, the sneakiest, most shameful thing he'd ever done, the memory of it still had the power to make him wince. Peaceful Lawrence with a green net fishing for sticklebacks in the stream. Shoes and socks neatly on the bank. Lawrence balancing on underwater rocks. Howard watching secretly from between the trees. If there's one thing he'd learned in life it was that his enemy cousin didn't like sudden noises. Howard watching. Lawrence rapt in his task, bent almost double, eyes squinting through the water. The thrill of fishy murder had concentrated his mind, lent, for the moment, his stubby body a touch of grace. He was chasing a school of fish. He tested the next rock with his right foot, he rested his weight upon it, left foot delicately up in the air, moving to . . . — and Howard, in the trees, let loose the loudest, shrillest, most hellish banshee scream. It hurt his throat but it hurt Charlie more, he hopped, he slipped, he waved his hands, the net dropped, floated slowly away, Howard heard him whimper as his left foot kicked up a big splash of water, as his right foot lost its hold, as he stumbled forward, grinding his right leg along the side of the rock, fell on his face, head went under, Howard could still taste along with him the retch of green sludge, as Lawrence smashed his chin against the underwater rock, splashed up, fell again, splashed up again, blood filling the gash in his chin, a tooth dripping out of one corner of his terrified mouth.

— Why are we doing this? I don't like dogs.

— Don't worry. I sort of own this place.

Howard pulled open the acrylic casing around the halogen bulb. He removed the rusted key hidden there, replaced the casing in its sockets. He fitted the key in the iron door. The lock resisted, turned. Howard eased open the door. He didn't feel drunk any more. They stepped inside.

Lights flickered on. Long corridor, oily handprints on walls. They walked along, legitimate interlopers, she sheltering under the safety of his arm, he touching each new surface with the palm of his free hand, the toilet door, the recreation room, a console of light switches. And, they turned, they walked up three concrete steps, he pushed open the heavy door, they stepped onto the factory floor.

The door closed behind them, an asthmatic's dying wheeze. The silence and the blackness were unnerving. The smells were reassuring. Workshop grease, oil, last year's sweat. It had always smelled like this.

— What are we doing here?

— This is the factory. Can you imagine it on fire?

He had never switched the lights on here, like Carol in the taxi he didn't even know where the switches were. He waited for their eyes to adjust themselves to the dark. The machines slowly showed their shapes. Big and elegant, standing in upright diagonal rows, his father's favoured children. Baskets beside them. Drums of polythene granules, raw plastic, along the far wall. Howard and Carol walked along avenues of machines. He laid his hands upon them, steel control plates, electronic buttons, computer fascia, glass-fronted clocks.

— They're big, she said. I'll give them that. What are they for?

An impulse. They were at the end of the middle row. He checked the machine in front of them, it was ready.

— Different things. I'll show you.

Howard pressed the button: a whir, a whine, the hum of gathered energy: the rush of raw plastic, metal moulding plates slid, met, hugged and locked, the consummation of shape: plates parted, the hot perfect form slipped down through the cooling chamber. And again and again, and again, production cycle procreation: mould, kiss, separate, expel: green toy goblins rolled down the chute into the basket. And again. He switched on a second machine. They listened to dozens of unbreakable combs formed in heat.

The noise was good. They walked on, to the dustiest, rustiest corner of the shop floor.

— This is called Museum Corner.

It had been Museum Corner for as long as he could remember. Squatter machines these, rusting metal, unconnected to electricity, abandoned prototypes for lost machines, which no one had ever been able to figure out any function for. And if the factory floor had any windows, if the lights were on, Howard would show her the indecipherable words inscribed beneath the dirt on the brass name plates. He ran his fingers along the words, tried to read them by touch.

The combs and goblins continued to be made. The comforting sound of manufacture followed Howard and Carol as they left the factory floor and went back into the light, down concrete steps, along the corridor.

— This used to be offices. It's a storeroom now. Sam's office was over there, I think. It became Freda's. My father was never here. He was long dead by the time the company moved here. That was Jackie's room. Jackie was the foreman, kept on with my father's work. Improved on it even. He developed a hot runner system, back in the mid 'seventies. And then he asked for a piece of the action and Freda fired him. He went to work for Harry Wiener after that. Then came Avi. He was a table-tennis wizard and a couple of other things besides. And you see there? Sort of around here, in the middle, that was where the chief salesmen operated from. Joe sat over there and Sid over there. It's where Charlie and me learned the business.

— I can still hear those machines. Shouldn't we switch them off? Or something?

There was something inside of him, like a thin gently pulsing cord, which was organic, pink, translucent, and tied. Memories craftily hid in every cranny of this place. Here he had spent a school holiday reading comic books in a revolving leather chair. Here he had spent his months with Joe the Poet, listening to him trying to sound enthusiastic on the telephone, watching him in spare moments scratching spidery half-rhymes on to a yellow notepad. Here he had learned how to apply spin to a table-tennis ball. Here a mystery assailant — it must have been Charlie (and curiously, he cannot separate Charlie from the Lawrence he had been back then) — had hit him over the head with a fibreglass rod. Here he had held back the tears of shame and rage as he declared to the gathered heads of his family that he would never join their company. Here he had hidden behind filing cabinets and watched Jackie's daughter Millie lift up her skirt and pull down her knickers and receive a shilling from Charlie (which had been tricked out of Howard's pocket in the first place). Here he had arrived for his first day of work. Here he had been made managing director. No. That was upstairs. In the offices in the New Building.

— Avi was on the make. He ran his own business. Used one of the machines to manufacture costume jewellery, which he sold to the workers for their wives. Freda sacked him in the end.

— I mean. Really. Shouldn't we switch off those machines? They're enough to wake the dead.

— Let's go upstairs.

Into the reception area, pushing through the wooden waist partition that was the only memento of Danny and Sam's first factory, up the stairs to the New Building. In this office he had conducted his first general meeting. In this toilet he had first noticed his hair loss. This was the telephone that had brought him the news that his wife had gone into early labour with their second child.

— Tell me about your family.

— Monks don't have families.

His voice sounded slightly more prissy than he had intended but it still reached that note of saintly self-satisfaction that so became him.

— I know you were never a monk.

— But . . .

— It's just nice sometimes to pretend. Did you think I was stupid?

— I never thought you were stupid.

His imaginary monastery disappeared. He had been planting the gardens with vines. He had built shelves for the library and had been filling them with illuminated manuscripts. Footsteps echoed and died around the cloisters.

— I've always wanted to seduce a monk. What does this place make anyway?

— Plastic things.

— I knew it'd be boring.

Here was the room that used to be his office. It had been, as far as he could tell, undisturbed. Papers where he left them on the desk. Window opened slightly to let out the smell of Charlie's cigars. On the wall, over the mahogany mantelpiece, between tiny industry trophies, was the white button mounted on cream card and framed in silver, with the signatures on the yellowed label below, *Danny Sam*, and underneath the neatly typed date, *January 7, 1940*. The first fruit of Danny's genius. The urea formaldehyde foundation of the family fortunes. Carol sat down behind the desk. Howard brought out the photograph again, held it down on the desk between them.

— That's my father. He married his brother's wife's sister when—

— No. *Your* family. You have children, don't you?

She was merciless.

— A boy and a girl. Ruthie is kind of wild. She's fifteen. And Simon is thirteen. He's, he's . . .

— What? What is he?

How could he tell her? It was the family secret. Their shame and in a curious way their pride.

— Why is he called Porky?

— What do you mean, why is he called Porky?

— It's a joke.

— I don't know. Porky? I don't know.

— Neither do I. You want to see Charlie's office?

Through the connecting door, past the shared executive washroom, into Charlie's realm. Howard was a phantom in this place. He made no sound as he walked. This is the world that was once accidentally his. A victim of birth, the captive prince of other people's ambitions.

— Wow. He really likes plastic, doesn't he?

— I suppose he does. It's his touch of mysticism.

Everything in here apart from the cocaine and cannabis in

the top drawer of Charlie's black polypropylene desk was plastic. The paintings, the stereo, the computer, the window, the furniture, even the lining on the walls, all of it was proudly, loyally, smugly plastic.

— I was in charge. Charlie was working against me. He was always working against me. I was losing the stomach for it but that's not why I did it.

— Did what?

— Let's get out of here. I don't like it here. It gives me the creeps.

— Why is it like this? Is the place going broke or what?

— What do you mean?

— If there was money you or this Charlie could afford to have everything nicer. Like wood. Or marble even. Plastic's tacky.

She sat at Charlie's desk. She rocked and then she revolved on Charlie's chair. She started to open the drawers of Charlie's desk and Howard had to fight down the territorial urge to stop her.

— Plastic isn't tacky.

— Of course plastic's tacky. Everyone knows that. No style. It's, vulgar. Look at this office. You trying to say this office isn't tacky?

— But that's only because Charlie's tacky. Plastic can be anything you want it to be and that's the whole point of it. With a good enough mould anything is possible. If Charlie had class then this office could be beautiful and it'd still be plastic.

— You mean it, don't you? Do you still work here or what?

— No. I don't work here any more.

— I hear a lot of businesses are closing down. That's what they say.

— I'm leaving it in good enough shape. People always need hair products, the rest will fall into place. I don't feel guilty about that.

— Who said anything about guilty? But answer me this one. If plastic's so clever why isn't everything made out of it?

— Like what?

— Like this. Old-fashioned paper. I found it with the drugs. It looks secret or something.

Charlie's desk drawers were open and Carol was holding something for Howard to take. He took it. Three paper pages bound together with a purple PVC spine clip. The initials of all the company directors except for his were on the top sheet.

LPC

A Report On The Death Of Mrs Buller
Classified

Overview

On 30 May, Mrs Marjorie Buller, of 42 Dover Road N16, was discovered dead because of an overdose of painkilling tablets and sleeping pills. The tablets had been prescribed for Mrs Buller's sciatica and arthritis. The pills were for Mrs Buller's insomnia. Mrs Buller was a 64-year-old widow, with a medical history. She lived off her state pension. The coroner recorded a verdict of Accidental Death, but entered into the record a reference to the pressure of 'incompetently directed business correspondence' as being a 'notable contributory factor' in 'Mrs Buller's confused and unhappy state of mind prior to the event of her death'. Among those items of business correspondence were several letters from the Accounts and Legal Departments of LPC Limited.

The Facts

Dover Soul Ltd, of 42 Buller Road, N16, is a customer of LPC, having bought three extruding machines in the previous calendar year. It is severely behind with its payments. Dover Soul Ltd was registered as a delinquent account on 21 February. The first contact Mrs Buller had with our company was when she wrote to us on 23 February complaining of an LPC van having broken down in Dover Road, N16.

No letters from the Accounts Department were sent to Dover Soul. Instead Dover Soul continues to be the recipient of Direct Mail campaigns inviting the aforesaid company to take advantage of lucrative special offers. Dover Soul is now the subject of a suit brought by LPC Ltd's Legal Affairs Department for settlement of its delinquent account. That suit is likely to prove unsuccessful because no letters from LPC Ltd's Accounts Department were received by the defendant. LPC Ltd's Distribution Department is responsible for having filled Dover Soul's order for two more extruding machines, despite the company having been delinquent in its earlier order.

The first letter Mrs Buller received from the Accounts Department was sent on 26 February. It began as follows: It may have escaped your attention that you have not yet paid for the machines you received on 3 October (see accompanying invoice). We wish to bring this matter to your attention and trust to a speedy settlement of this account.

The second letter Mrs Buller received was sent on 26 March. It began as follows: We are sorry to have to remind you of our previous letter of 26 February inst. requesting a prompt settlement of your account. According to our records the account remains open, thereby accruing a 6% interest charge for late payment. Your account now stands at £18,408.62. We will be happy to waive the interest charge if the account is settled within ten days. If you have settled your account within the previous seven days please disregard this letter.

The third letter Mrs Buller received was sent on 23 April. It began as follows: It saddens us that a previously valued customer should place his or her credit rating in such jeopardy. Please see enclosed invoice detailing interest charges accruing to your account. We regret that the matter is now in the hands of our company solicitor. If you have settled your account within the previous seven days please disregard this letter.

After receiving this letter Mrs Buller wrote to us. That letter was fielded and responded to by LPC Ltd's Customer Relations Department. Nikki Giles, Acting Director of Customer Relations, wrote speedily back (17 May) assuring Mrs Buller of the company's sympathy, apologising to her in the event of any error having been made.

On 24 May, Dover Soul Ltd was sent a letter from our

After Sales sub-Department. It congratulated Dover Soul Ltd, for making a wise choice in investing for the future with LPC Ltd, and invited an executive representative of Dover Soul Ltd to complete the enclosed Customer Satisfaction Questionnaire.

On 28 May, Mrs Buller was sent a letter from our Legal Affairs Department. It warned her that her delinquent account had now been closed and the company was consequently seeking legal redress to recover its lost payment and expenses and the interest accruing upon these. The letter ended as follows: The full thrust of the law will be used against you.

On 29 May, Mrs Buller called a late-night phone-in programme (*Nightflight*). That particular episode of the programme was concentrating on medical and emotional problems with special guest Doctor Lottie representing *Teen* magazine. It is believed that Mrs Buller aroused the impatience of Doctor Lottie by giving a rambling account of what she termed her 'persecution' by our company. When she revealed the company's name, the 7-second delay was brought into effect to avoid possible legal action and Mrs Buller's call was terminated. Mrs Buller was discovered dead the next morning by her daughter Deirdre who had visited on the occasion of her mother's 64th birthday.

Conclusions

- The authors of this Report inescapably conclude that there is an inefficiency of communication between different departments of LPC Ltd.
- They deeply regret the death of Mrs Buller and any part this inefficiency may have played in contributing towards her demise.
- They suggest that it might be prudent to change the name of one or more of the departments mentioned in

this Report. It is believed that this would be strategically beneficial for internal morale as well as sending the right kind of signals out to our customers and competitors.

- Furthermore, the authors of this Report feel bound to recommend most stringently that a system of checks and balances be put into place and that a Committee be formed to deliver a Report into possible restructuring of the Company.

- Furthermore, the responsibility of the Managing Director must be questioned. It is believed that it is his presidency over the company which is largely responsible for the inefficiency of communication between departments.

- Reluctantly, the authors of this report must conclude that it is inappropriate for Howard Levy to continue as Managing Director of this company.

When he had finished reading, Howard put the Report away in his jacket pocket and led Carol (her nostrils dusted with white powder) into the washroom he used to share with his cousin — Charlie the usurper, the plotter, the commissioner of reports. He switched on the light, together they examined his reflection in the medicine cabinet mirror.

— Do you want to fuck me here? Is that why we're here? Is this some *thing*?

Angular face, long thin nose, high brow, flat cheekbones. Lines around the corners of his mouth and eyes which — it may have been the light, optical trick, he couldn't be sure — seemed deeper than the last time he looked. But the hair. He couldn't see it in the mirror; he tried, rolling his eyes up, bowing his head down. But he could feel it. Fingers first tentative then assured, stroking the crown of his head and finding only short growing hair, no scalp, no baldness.

— I think Charlie keeps some grease here, his emergency stock.

Howard opened the medicine cabinet. A white PVC tub, plump with orange gel, polyvinyl pyrolidone.

— Here. Let me.

She held out her hands, palm upwards. Howard dipped his fingers into the tub and wiped some of the orange goo into her hands, watched her roll it, warm it, stretch it. He crouched down, chin against the rim of the sink, and felt entirely at peace as she rubbed the gel hard into his hair and shaped it with her fingers and laughed. Every now and again she would make him rise to inspect the work she had done. Hair bristled up to make it spiky. A fringe down over his forehead, flattened with the balls of her hands. She forced it all back, severe, a killer's cut. She gave him a centre parting, a parting on each side, no parting at all, she flicked the hair up at the ears so he looked like a mass murderer pretending to be a silent movie heroine. Inside, the company man shrivelled and all the others yawned. They stretched and flexed, announced, with nerve-ending jabs, their hunger for adventure. He took out the photograph again, perched it on the basin, examined for what he hoped was the final time the mad eyes of his father. Later, tomorrow, the day after, whenever, he would deliver the photograph to his enemy's

95

place. Pass on the family curse and blessing. A pink translucent cord trembled, shimmered, snapped. Carol gave him a rocker's splendid quiff. The past was dead. The future would be, as the old slogan said, like plastic.

9

LIGHTS WERE ON in the windows at Carol's place. *That means*, she said, *that Mickey is in and waiting. Best not push it* ... Howard watched her go in. He waited, as if something remarkable might be about to happen, until the lights had been switched off; then he took the cab on to Charlie's place.

Howard had expected Charlie's building to have better protection. He followed a Range Rover through the security gate into the car park and then he followed its poodle-haired driver into the building and the uniformed guard didn't even bother to look up from his desk. Straight on Howard went, perhaps he was invisible still, along the marbled floor, past hideous paintings of technological monsters and a water fountain lit to make the water look like light, straight into the waiting elevator, up to the penthouse floor. Then along a grey carpet to Charlie's flat. Perhaps, Howard thought, he should allow this to become a meeting. Press Charlie's buzzer. Wait for him to open the door, bleary-eyed. *You want to see me? Well here I am. Take this photo of our fathers and leave me be, my future is like plastic, I'm running. I'm running properly now, out of town, to where nobody knows not even me ... And you think I killed Mrs Buller? Well think again, we all can be proud of that one* ... No. It would be better, more unsettling this way. Howard kissed his father goodbye and slipped the photograph inside an envelope taken from one of the secretaries' desks at the factory. He used the apartment door for support as his left hand awkwardly wrote Charlie's name upon the front. He pushed the envelope through the letter box slot. It balanced there for a moment, waved, and dropped.

The photograph waited in darkness on a rubber welcome

mat and Howard was gone. Howard was running. Renegade Howard was in a taxi cab, on the street, at a train station. He was buying his ticket, he was surrounding himself with the things that he decided to need for his journey, cigarettes and beer and a book of poetry to read. We can't reach out to Howard any more, let him run, we're inside a roughly addressed envelope; leave Howard to his train ride, and trust he will come back to us — the train picks up sluggish speed, the carriage lights flicker, someone's beer can topples, splashes, rolls, already we can no longer read the station name in the blur of signs rushing past — later.

THREE

THE GOOD SON

**Did you always think the Renaissance
was for other people? That only rich
people could afford the Old Masters?
That Michelangelo's Moses was just
too BIG to display in your home?
Think again.**

After years of scientific research **Classic Art Club**™
brings you an exclusive new SELF-CLEANING range
of Miniaturised plastic PRESTIGE ART FOR THE
HOME. Put Moses on the Mantelpiece. Hang Botticelli
in the Bathroom. Have The Sistine On YOUR Ceiling!
AN EXCITING NEW DEVELOPMENT. For YOUR
FREE GIFT simply sign the coupon below and send
it FREE OF CHARGE to **Classic Art Club**™. You
are under No Obligation to buy. ENJOY your FREE
Last Supper in 3-D!! in the Comfort Of Your Own
Home and if you are not fully satisfied Send It Back
within ten days. You will receive your COMPLIMEN-
TARY LAVISHLY ILLUSTRATED copy of *Classic
Club*™ *News* EVERY MONTH, from which you will
be able to buy EXCLUSIVE AUTHENTIC REPRO-
DUCTIONS AT SUBSTANTIAL DISCOUNTS. You
may cancel your membership At Any Time.*

Classic Art Club™ Members agree to make six purchases in their first year of membership.

Advertisement in the *Sunday E —— magazine*, 16 March 199–

CHARLIE WOKE UP farting. A deep blast like the sustained note of a bass saxophone, it began somewhere in the curried belly of Charlie's sleep, and climaxed, elegantly, with a trumpet trill, a brief promise of perfect silence, then one high pitched after-toot, and its inadvertent creator was awake and smiling at the ceiling. He enjoyed it so much he tried another. A quicker one this time, shrill, piercing, almost brutal, a warning call from a high-flying hawk.

The third fart was closer to the splendour of the first, a church bell sheet-ripper of a fart, sombre, magnificent. It would have befitted a monarch's funeral, the slow winding cortège through black-laced streets, the weeping widow leaning for support on the narrow shoulder of her dead husband's sinister private secretary, the Archbishop with fading eyesight waiting in the cathedral, his slack face hiding his nerves behind an expression of infinite understanding, arms behind his back, funeral address picked out in braille hidden in the wrist ruff of his robes, and Charlie's grand fart thundering through it all, reminding the mourning city of terror and pity. Charlie followed that one with a saucy little bleat, a naughty French girl of a fart, eyes open wide, hand over startled mouth, *Oh, Monsieur Shar-lee!*, limbs so innocently arranged but somehow the neatly sewn hem of her chaste blue skirt has rolled half-way up her black stockinged thigh . . .

Charlie trumped. Charlie guffed. A sphincter symphony with Charlie the jaw-clenched eyes-screwed conductor battling a rebellious orchestra. His ideal was a round roar, a simple perfect explosion; his buttocks would part (and sometimes he would slide a finger between them to feel that ideal airy expulsion) and out would roll a gassy bubble of pure manly sound. Charlie grinned proudly each time he achieved perfection, inclined his head towards his imagined brass section. But some popped out as insipid duck-quacks, or worse, a thin parp, a dribbly little sound. And it didn't matter how hard Charlie heaved with his muscles, the forced farts were often disappointingly shrill, the best were those he was patient with, babies he gave time and love to, feeling them strengthen inside him as they grew warm to make their impressions upon the world, and with the very best — and this is when Charlie felt upon him the shivery breath of the sublime — it wasn't he who was producing these happy farts, instead he felt the farts farting *him*.

Charlie launched fusillades of farts. He steamed, he belched, he belted. He roared, he bellowed, he blasted; he whimpered. He hummed a low tune slowly along with the best of them. He chuddered and shuddered and burped. He was a volcano that rumbled. He was a volcano that erupted. He eructed. He let off. He chuffed. He odorised the room with the gay scent of last night's vindaloo. He crepitated with joy. He gurgled with pleasure. He let rip. He gurked.

He was nearly spent when a dim recent memory started to become a little brighter. A fond little pop of a fart made her image more clear. A girl with blonde hair and green eyes and a lascivious taste for hot Indian food. They had met at the club or else it was earlier at somebody's drinks party. She was a doctor. She had a quick broad smile and her name was Jane and she had the most beautiful calf muscles he had ever seen. They had danced and they had eaten and he had told her some lies about himself and some truths too. Charlie slowly opened his eyes. With great care he turned his head. She was there, lying beside him, Jane with the calf muscles. Her body was unnaturally rigid, her wide green eyes were staring at him. For a moment there was silence. He tried a smile. Then a final fart peeped out, an airy unrequested encore, and her body flinched and the muscles at the corners of her mouth twitched and a vein pulsed at her temple and an expression of utter horror polluted her pretty face.

— Ah, said Charlie. Uhm. Ah.

When Charlie woke for a second time Dr Jane was gone. There was an empty space beside him. No calf muscles or smile or leather bag of medical tricks.

— Fuck, said Charlie.

He removed himself from the bed. It was an effort of will to remind his body that it still belonged to him. He stood, slapped his belly, watched it ripple, coughed, farted, rolled his head around and moved his shoulders up and down. His arms hurt. But then so did everything else. He spoke in a Vincent Price voice to kill the silence.

— Welcome, my foolish young friend, to Babelcunt Castle. I am master in this house . . . Shit.

To the bathroom. Radio on, electric guitar music, loud. He pissed like a whale. Then into the shower, jet set to Maximum Massage. He yelled full-throated football fan *YES!*es as the water powered against grateful flesh. Rolling Stones on the tape machine in the kitchen as he made himself a mug of Multi-Strength instant coffee. Coffee gulped down and it was back to the bathroom, where a sexy-voiced advert for a hernia hospital kept Charlie company for a quick shave before he put his contact lenses in. More guitar music, primitive, happy, as he greased his hair into a pompadour. And, with his lenses in but with his eyelids shut (he was not yet strong enough to brave the gaps in the steam on the wall mirror), he smoothed some of the ruin away from his face with a French lotion left behind by Rocky Sporgente's girlfriend.

Breakfast was a microwaved pair of Heat 'n' Eat Chicken Kievs eaten as the Small Faces serenaded. He left cash for the cleaning lady underneath the ashtray in the dining room and he finished the joint he found there and the two glasses of wine beside it, and the remainder of the crossword puzzle in yesterday's *Financial Times*. It was time to dress. His business gear. Houndstooth checked suit. Open-necked sky blue shirt. Silk underwear that matched, a fastidious touch, the handkerchief in his breast pocket. It was while he was searching for his panama hat that he saw the envelope lying on the welcome mat.

It took a while to focus on it. Took longer to prise the envelope away from the long rubber pimples of the mat. He held it. It was a manila envelope, like the ones they used at the company. He shook it to hear if it rattled. Waited for the words to shimmer into focus. His name on the front in a moron's roughly printed writing. What would be inside? A death threat maybe, some small-dicked cuckold's pathetic attempt to unnerve, de-erect Don Juan Charlie. It wasn't a note, felt thicker than that, a party invitation? Too thin. He opened the envelope, struggled for a moment to slide the thing out. It was a photograph, creased, crinkled — a blackmailer?

— What the fuck?

Charlie hadn't seen this photograph before. Brittle, old, black-and-white, his father and his uncle in workshop clobber in the morning of their triumph, showing off their first unbreakable combs.

Charlie Levy walked like a gangster. His first wife told him that once, in the kitchen at their place in Highgate, when the walls had been suddenly redecorated with ketchup, broken glass around their feet. She'd meant it as an insult and if she'd known him at all she'd have come up with something better. He walked like a gangster, a cocky fucker, out of his apartment, striding like Cagney to the elevator, humming along with the muzak, hat tipped to his jaunty reflection. He tipped his hat again, when he was downstairs at the reception desk (surrounded by acrylic paintings of machine elements on the walls — part of Charlie's own collection, lent to the building, rent appropriately adjusted as a result), with a uniformed tough in front of him.

— Morning, Terry. Mark here yet?

— Not yet Charlie.

Charlie held the envelope out for the tough to take. The tough kept his attention on the console of security screens.

— Who sent this? A mongoloid?

— Couldn't say.

In the middle of the security screens the doormen had smuggled in a television, cunningly disguised in black and white, fooling nobody.

— Could be someone trying to hide their handwriting. What do you think?

Terry Tough looked reluctantly up — at Charlie, at the envelope — then he shrugged his shoulders, allowed out a cold, gold-toothed smile, and looked down at the screens again.

— Probably so, Charlie. Couldn't say.

A bell sounded. Terry checked a subsidiary image, flicked the switch that operated the gate. Charlie craned his neck but he couldn't recognise the programme Terry was watching. This was a bad start to the day. Even the writing on the envelope troubled him. Like valentines sent to the tight-fannied office siren, or an anonymous tip-off to the Inland Revenue about a competitor's sneaky ways. Whoever it was probably wrote it with his left hand.

— Mongoloids shouldn't be allowed to just sneak in. They've all got schlongs like bulls' pizzles I understand. We could have been raped in our sleep. What do we pay you, you chaps, *for*?

— You can make a complaint. The Board will look into it. Are you making a complaint?

Everything's a Board these days. Wherever you turn, fucking planks of wood. Mark skidded the car into the car park — puddles splashing, exhaust fumes puffing lively ghost shapes into the air — and Charlie let the envelope drop to the floor. Terry Tough could pick it up, give him something to do for his money. Charlie put the photograph away in his back trouser pocket. He walked out into the car park, stood well back as he waited for his driver to finish performing his fancy splashing turn. He wasn't used to seeing his father alive and young. It was unsettling. It — and this was hard to admit — *meant* something. Charlie sniffed the air. He spat out the ensuing phlegm. The subject was closed.

2

CHARLIE'S CAR DROVE fast out of London. Wednesday had always been road day for Charlie except now he was on the road to buy rather than sell. Charlie's car was an old car, a 1971 Scimitar with a silvery blue fibreglass body, its curved handsome lines the gift of a designer with oil on his hands and sex on his mind. The engine sang, the dashboard was plastic, the seats shiny and leather-effect. Charlie's car was steered one-handed by Charlie's driver, a hard-faced pretty boy with long black hair tied in a pony tail. Charlie slouched in the front passenger seat. His jacket was bunched up airily over his big shoulders. His bare feet were splayed against the windscreen. The brim of Charlie's panama hat was lowered over Charlie's eyes. He had a financial newsletter open and ignored on his lap and a hip flask of brandy tilted towards his mouth. The boy kept fiddling with the graphic equaliser on the stereo. Something ugly was crashing through the car interior, a melancholy male croon over a juddering electric bass guitar and acoustic cello, which the boy sang softly along to.

— There's a thumb-jockey.

Charlie pushed up his hat and looked out of the window. He saw a broken-waisted sign advertising a historic quarry half a mile ahead. He saw straggly cows in a straggly field trying to munch and shit their way to contentment. Yellow wildflowers like jaundiced fingers were reaching out of the grass verge as if trying to warn the cars of something horrible ahead.

— I never could stand the countryside. I do like the seaside though, always have.

— I said there's a thumb-jockey. You want I should pick him up?

— What's that?

— You want I should pick him up?

A bearded man ahead dressed all in brown, down on his luck maybe or just some sex monster out on a midweek adventure, was holding a cardboard sign that read CARDIFF PLEASE M4. There was an overnight bag between his feet. The cuffs of his trousers didn't speak to his shoes.

— This music is deafening. Turn it off, will you.

Mark scowled at him but Charlie was hiding his eyes under his panama again. He screwed his hip flask shut, returned it to his breast pocket, and stroked his fingers along the skin of the Australian snake that ribboned his panama. The boy twisted the volume all the way up, grinned the sort of grin that girls must like, muttered an aggressive apology, and cut the power to the stereo.

— Stop the car. We'll give the gentleman a ride.

Charlie's car veered suddenly from the outside to the inside lane — making a Mini in the middle lane brake abruptly and shudder and cry — onto the pebbly verge where it halted flamboyantly with a screech of rubber, a brief skid, a display of sparks. Charlie opened his door and stuck out his face and pulled the hitchhiker towards them with a wave of his hand.

— God. Look at the state of that.

The brown man ran towards them in a high-stepping lope, carrying his overnight bag as if it was a pole he would use to vault himself over the car. Charlie slid himself out of the car. The brown man separated his brown moustache and ginger-brown beard. Charlie took that for an attempt at a smile.

— Get in, son. Make yourself comfortable.

The brown man climbed into the back seat. He held his overnight bag primly on his lap. The car shot out in a quick diagonal back into the outside lane. The force of the acceleration tossed the brown man about and he smiled at the fun of it. Charlie turned to examine him.

— It's good for the circulation. I'm Charlie and this is Mark.

— And I am Mark too, said the brown man in a flat foreign accent.

— Mark Two? You of German design?

— I am from Switzerland in fact. Student of sociology upon my holiday vacations.

— Been here a while?

— A long while. Much to my surprise and original discomfort I discover unsurpassable difficulty with maps. It gets so that I find I am unable to leave and then it gets so that I forget about leaving. Now I explore the Celtic tradition.

— That's nice. Let's smoke some reefer.

Charlie pulled a ready-made joint from an inside pocket of his jacket. Mark 1 jabbed an elbow down hard on a button to make all the windows go down. They bobbed about in the gale.

— He doesn't approve, Charlie shouted to their passenger as he lit up with a wind-proof lighter. Prissy, I think.

So, apparently, was Mark 2. He took two drags on the reefer for etiquette's sake and refused any more. This disappointed Charlie. He liked to share his pleasures. Mark 1 bullied the car faster and yelled some short unhearable words in delight at the speed and he turned on the stereo again and started to sing along again. Charlie smoked his reefer and held on to his panama and shut his eyes and started to think, in wide shiny images that filled his head and his future, about share certificates and percentages and the joyous necessity of forgetting all about the past.

Charlie tossed the finished reefer out of the window past Mark 2's worried bearded face and turned to watch it skipping and sparking down the road behind them. Mark 2 had hair climbing over his cheeks nearly to his eyes. Charlie was not normally fastidious about the eccentricities of bodies but he made an exception for this one. His driver lowered the volume on the stereo and closed all the windows.

They travelled fast. They made service stations disappear, and motorway cafes, and roadside signs vainly indicating points of local interest, and rape-seed fields, clumps of trees, and gangs of workmen playing about with traffic cones. Sometimes passengers in other cars would turn to look at them but before their mouths had finished their goldfish gapes Charlie's car had already zipped past. Mark 1 leaned forward, only losing his stillness when he needed to honk and curse a slower-moving car out of their way. Mark 2 fetched a coverless guidebook out of his bag and settled in to read all about the Celtic mysteries of Wales. Charlie shut his eyes. There was reassurance in speed.

Charlie roused himself. He had been napping. He wiped the corners of his mouth dry and caught his driver's eye and raised an eyebrow.

— Another half hour, said Mark.

Charlie twisted around to address their passenger.

— What, he said, would you like more than anything?

Mark 2 goggled at him for a moment.

— Like? What is it I would like?

— That's it. What would be the thing, the little magic special thing, to make your life complete? Mystery ingredient X. Name it. A battery-operated razor blade? A seat on a plane to Geneva or Zurich or whatever? A busty little fraulein to sit on your face? Tell me. I'd be interested to know.

— Indeed there are many things I would like to have including the ones that you say but I think the thing above all that would make me truly happy is the capacity to draw pretty pictures.

— Pretty pictures.

— Yes. Pictures of spires of ancient churches and wild animals with powerful bodies and the faces of girls that I love. With charcoal or a soft pencil perhaps.

— And what would you give?

— Give?

— In return. What would you give to have the capacity to draw pretty pictures?

— I do not know what I have to give. I would give money for sure but I do not have much money.

— Your shoes?

Both Charlie and Mark 2 looked at Mark 2's shoes. Old sturdy shoes. Brown suede uppers. Deep strong soles. He had slept a lot in those shoes, and each morning they forgave him.

— I am afraid this is the only pair of shoes I own. I think I would be not so willing to give up these shoes.

— Even if you were able to draw pretty pictures as a result?

— It is a tough question for sure. Maybe I would give up my shoes if it meant I could draw pretty pictures. But where is the teacher who would demand my shoes in order to show me how to draw? I do not think there is such a teacher.

— And what would you say if I promised you the capacity to draw? If I said to you that I could *guarantee* you would learn how to produce pretty pictures in a matter of minutes and all that you would have to do — call it an enabling fee if you like — is to give me those shoes you've got on.

— I think I would say to you no thanks. I think I would then laugh because you would be making a joke.

Charlie rested a hand on Mark 2's brown-jacketed arm. He lowered his voice as if they alone shared some important secrets in a hostile world.

— I can give you that guarantee.

— I do not think so.

Charlie pulled away and opened the glove compartment. He brushed aside piles of documents and business cards and a portfolio of Classic Club samples. He lifted up a flat, grey-lidded plastic box.

— This is a drawing machine.

Mark 2 looked doubtful.

— Battery-operated, computer driven. I could show you how it worked but I would want your shoes.

Charlie replaced the box. He closed the glove compartment door.

— With computers to draw? I have heard of such things.

— It doesn't use charcoal or soft pencils.

— Instead — what?

— A hard pencil, an electronic pencil, stroke it over the screen and the image appears thin or broad depending on the pressure. You need to correct your image? You want a straighter line? Choose it. Position your cursor and press a key. Change the angle? Press a key. A curved line? Another key. You can draw free-hand or you can cheat a little — use the onboard camera to make an image of the pretty thing you're looking at — then trace the outlines, store it in your image library if you want, add colours, print it out, a pretty picture. Zut alors, the thing is done.

— I would like to see how such a machine works.

— I would like to have your shoes.

110

— For my shoes you would give me this machine?

— For your shoes I would give you this machine.

A pause, a stand-off, which Charlie broke with a new, speedy kind of patter. Selling, like everything else, was a rhythm thing. Rhythm and confidence.

— Look. Take a look. You see this?

Reaching into the glove compartment again, into the portfolio of Classic Art Club™ samples. Perfect. Just the right size. Hard not to believe they didn't come out of this box. The suitable ones he stacked beneath the dead computer, pulled them out one by one to show to Mark 2.

— You see? You see? Venus by Botticelli. Gorgeous or what? Look at the finish, that's extra, special process, need heavy duty paper, I can throw that in for you, but take a look, yes? Better than the real thing. Took, guess, how long? Seven fucking minutes, excuse me, from start to finish. Along the way I made a few mistakes, used the quick erase, in the States we market it as the E-Z wipe. Or how about this, I can see you're of a spiritual bent, Dürer, lovely, two hands praying. Both of these camera-assisted of course but I have to tell you this is the great leap forward. Just think of it, none of this poncey I'm-an-artist shit. We're talking cultural democracy. Art for all. With this lovely little gizmo everyone's doing it.

Charlie flapped open the computer box, gave the customer a quick flash of keyboard and screen, then snapped the lid shut again.

Mark 2 looked at his shoes and then he looked at Charlie and then he looked at the artistic riches in his future and then he looked at his shoes again.

— I think I will give you my shoes.

He leaned down and started to pick apart his frayed shoe laces.

Charlie felt desperately sad. He reached inside the glove compartment and paused.

— I would have to tell you. It doesn't work inside electro-magnetic fields. That's a small drawback. It won't work in cars or planes. No problem in trains though.

— It doesn't work in cars?

— Not in cars no, nor aeroplanes, nor next to computers or microwaves. And dishwashers too can cause some interference.

— You are saying perhaps that you cannot demonstrate this machine inside the car while we are in motion?

— On the nose. Couldn't do that. And I'm afraid we're in a rush son. Can't afford the time to stop and demonstrate it to you. That's the only problem facing us.

— Then I would have to trust you? I would have to believe in the words you are telling me.

— That's right. You would, if you're not of an entirely trusting nature, have to take a chance.

— I do not think I can afford to take this kind of chance.

— That's your decision.

— I am not a rich man. I cannot afford easily to buy new shoes.

— Fair enough.

— Without the shoes I already have my feet would be sad.

— I'm sure they would be. Believe me son it's generally not a good policy to give away your shoes to strangers just because they've given you some fancy patter. No proof? Hang on to your shoes, that's the principle.

Charlie grabbed his customer's arm, shone the light of conviction upon him.

— But maybe, now and again, once every so often, someone offers you an opportunity from out of the blue. It might be something you've always wanted, it might be something you've never thought of before but is going to change your life. It might be an opportunity that's never going to be repeated. Faint heart ne'er won fair lady.

— May I see inside this box, please?

— No.

— Excuse me, why?

— I'm sorry, son, but the answer's no. It will only confuse the issue. I'm not going to stop the car so all the slick-looking gadgetry in the world won't change the issue. It's up to you. Here's a grey box that *might* be what I claim it is or it might be just a clapped-out lap-top. And there on your feet are your shoes. It's up to you. You have all the information you need to make a decision. You don't need anything more. Believe your uncle Charlie.

— I do not think I can afford to do that.

Charlie closed the glove compartment. He slouched back in his seat again. He winked at his driver.

— You are a rich man in a funny car and I am thinking maybe you become a rich man because you take advantage of people like me.

Charlie tipped his panama over his eyes and yawned.

— But then I am thinking too that here is a machine that will allow me to draw and I am thinking that you might be a rich man with a peculiar nature and that you tell the truth about your grey machine. I do not know what to think. I do not want to lose my shoes. I should like to be able to draw.

An agony danced on Mark 2's face. Veins pulsed in his eyelids. He pulled at his beard with long graceful fingers that itched to draw pretty pictures. He leaned down, pulled off his beaten suede shoes, held them up in the air, gave them a last sorrowful look.

— Please. I make my decision. Here are my shoes.

— No dice. Deal's off.

— I believe your words. I am happy to take your machine for my shoes.

— Sorry.

— Please. Take my shoes. Here. Take them. They are yours now.

He leaned forward. He placed his shoes on Charlie's lap. He patted them down as flat as they would go.

— I do not need to look inside your lap-top box. I trust your words. I am grateful. I shall draw cathedral spires and the faces of pretty girls. You shall look after my shoes and perhaps brush them when you can and there is a spray you can buy in Switzerland and I am certain in your country too. Please. Instruct your driver to stop the car and leave me at the side of the road and give me in exchange your drawing machine and if you cheat me I do not care. I do not believe you cheat me.

Mark the driver switched off the stereo.

— We're off the motorway in a quarter mile. There's a tuck-and-fuck just ahead. We'll drop the Nazi there.

They jerked off the road into the forecourt of a pink plastic place with a little pink hotel and little pink restaurants and pink toilets and pink shops. Charlie stepped out and held the door open for their passenger.

— Believe me son. Keep your shoes. You want to draw? Here's a tenner. Buy yourself some crayons.

Mark 2 climbed slowly out, with his shoes in one hand and his overnight bag in the other and the ten pound note flapping in his mouth. He held out his shoes towards the car and looked hopefully back inside as if he might be about to receive his grey drawing machine after all. Charlie tried to smile reassuringly at him but he felt suddenly too tired to do that.

Charlie's car pulled back onto the motorway, Charlie winked at his driver and his driver looked away.

THERE HAD BEEN an agony aunt in Notting Hill who made him camomile tea and shared a reefer with him while they completed their transaction and talked about her Jewishness as country music jauntily played. Then an Australian quartet of three brothers and a nephew over-filling a hotel room in Bloomsbury, outdoor Jews these, with broad chests and cancerous faces. In Holland Park Charlie had drunk whisky with an actor half-familiar from a television ad campaign of ten years before and then one awful cab ride to Kensington to meet the actor's impresario brother, bow-tied, actorly, cheerful. He had dealt with a drug dealer Jew who lived in a ramshackle Hackney flat where, as they breathed in the bad air soured by fags and dope and babies, Charlie bought what he had come for and two illicit bags besides. There had been a university professor in Manchester, a pianist in Brighton, and an old lady, Saul's widow, dying in a rest home in Dorking. A film industry Jew who had redesigned his body in the gym was the most intransigent and up till now a journalist in Cardiff had been the least. Religious Jews, lost Jews, rich Jews, hard-working Jews, waster Jews, society Jews, underclass Jews, petit bourgeois Jews, Jews who have no idea they are Jews at all: and what do they all have in common? Not blood types or shapes of noses or attitudes to money or the state of Israel. No. There are two things that bind these people together, that force them, unknowingly, to belong to the same category. One is sex and the other is property: the fact that whether they know it or not these Jews owed their lives to the linking in darkness of the loins of Jack Levy and Sarah Levy née Bremen; and their temporary ownership of a piece of heavy, cream, bonded paper, orange eagle crest at the top, orange oceany border, ornate black writing in the middle which declared a share of the ownership of LPC.

Charlie has met them all. As Lawrence Levy, the long-lost cousin with a tremor in his voice and tears in his eyes. As Mr Mann, stern liquidator, who works, he hints, for a sinister government department. As Charlie Levy, with a bucket of money and a bag full of jokes taking something meaningless and sticky off their hands. Charlie Levy owns the company now. Turn in your graves you stale saintly fuckers, the future belongs to Charlie.

He has made love to Rocky Sporgente's girlfriend on a bed of share certificates. Tenderly, gently, no biting or beating, as close to innocent fornication as her boredom would allow, they made a sandwich of her pale flesh, he buttering down upon her and, below her, rustling like lettuce, sixty-seven per cent of LPC.

Already he was in full control. He was Managing Director (thank *you*, Howard), and king shareholder. He controlled the Board, he could make every decision, unchallenged, unchallengeable. What he could not do was change the status of the company. He could not sell it and he could not close it down and he could not alter its charter. Dead people had decreed that the meaning of the company's life was the manufacture of plastic moulding machinery, and/or plastic products. Live people could not change the minds of dead people unless they owned more than seventy-five per cent of the company shares.

This one's name was Georgia. Georgia produced television programmes. She was small and slim and had dark blonde hair and a graceful way of wearing loose mannish clothes. She had a high brow and sensitive features that didn't take the trouble to smile and this charmed Charlie. They were related to each other through a turncoat cousin who had gone by the name of Paul Owen, but Charlie hadn't come with photograph albums and family trees. They drank tea from brown polystyrene cups in the canteen of the building where she worked.

She was pretty and busy and she gave him what he came for without resistance. He tried to spin things out but she signed the forms he put in front of her and slid a large BBC envelope across the table towards him and waited to receive the money in exchange. It wasn't much money (he would happily have paid twice as much and still know he'd got a bargain) but Charlie was beginning to feel cheated. He had come a long way for this and didn't care for it to pass without incident. He took his time to open the envelope. Inside was a share certificate, cream bonded paper, handsome orange and black design with the orange eagle at the top. One per cent of the company. Charlie breathed heavily, out of choice, and he wasn't entirely sure why.

— It looks in order.

— It is. I'm sorry you had to come all this way.

She was bored by the procedure and entirely without curiosity about him and this irked Charlie.

— Do you happen to know a good hotel?

— I thought you said you were going straight back to London?

— Did I?

— We're editing upstairs. I have to go.

— Your job must be fascinating. It's a nice town you've got here (were you born here? have you always lived here?). I was thinking maybe of spending a few days, making a holiday of it. It's lovely. I'm smitten. Roman, isn't it?

— I'm not much of a historian. Look —

— The street names make me giggle. Black Boys' Lane goes into White Ladies' Hill. I hear there's some stunning countryside outside.

She was irritated enough by him to show him what she saw, a metropolitan man with something amiss in his lungs who was taking something meaningless and sticky off her hands.

— Look. Couldn't we finish this? I don't have the time to chat about street names and Romans. I don't know about hotels. The Grand is the fanciest. It's on the Crescent, you can get directions in the lobby outside. You gave me the price on the phone, didn't you? I'm happy with it. A cheque will be fine.

— Of course. A cheque. Made out to . . .?

— Owen. Georgia Owen.

— Quite. I'll do it now. Your father's name was Paul, wasn't it?

— Did you know him?

— A very eminent man, in his field.

Charlie fetched out his bank book. Breathing heavily, he began slowly to fill out the cheque. He considered for a moment telling her about her father, Rebecca's son from her first husband, perhaps legitimate, perhaps not. He considered telling her that Saul Cohen had only become Paul Owen shortly before he married. Charlie decided against that as an unnecessary and unprofitable mischief. (In the same way he resisted the desire to build a hand bridge over the twelve inches of formica and cigarette air that separated them, to dip his fingers in the smooth hollow of her neck, to find her shoulders and arms and breasts under the striped shirt she wore, to try his mouth against her pink unpainted lips that reminded him, alarmingly, of Freda's.) He dallied over the cheque, made a deliberate mistake with the date, smiled, frowned, ripped it out of the book, sighed, tore it into small pieces and dropped them one by one into his pocket, and began again.

Georgia nodded at a cabal of men in suits at a nearby table. She leaned down, lifted a pack of cigarettes to the level of the table, quickly realised the encouragement this would give him, and returned it to its hiding place in her black saddle bag.

— I'm glad to be getting this over with, she said, desperately watching his pen scratch out laborious figures. It's been a nuisance.

— Oh? said Charlie. A nuisance? In what way a nuisance?

— People I don't know bothering me. Making enquiries. What is this company? You said it was liquidating, didn't you? I told them I'd sold my share but they keep ringing back. I've had to stop answering my phone.

— Must have been dreadful, Charlie said, wheezing sympathetically and hoping his grin didn't show. Did they leave names?

— Leslie something? Had a sort of wheedling voice.

Charlie lifted the cheque to his face as if he was very shortsighted and he puffed the writing dry and passed it slowly over to Georgia. She folded it, put it away in her bag close to where her cigarettes must have been. She nodded slightly to indicate their meeting was over. Charlie no longer had any desire to prolong it. There were things to do.

— Very grateful, he said. Lovely to meet you. Thank you.

118

Georgia shrugged her shoulders as she got up to her feet. She hoisted on her saddle bag. They shook hands. Charlie smiled. Georgia didn't.

— Yes, she said.

He watched her stride to the exit door past hairy technicians hunched over their lunches. He had got what he came for and something else had been confirmed for him besides, and as he followed Georgia out of the staff canteen (his gaze fixed on her tanned bare ankles above black brogues) he was touched, very lightly, as if the gentle old finger of God was tapping at his shoulder, by a dangerous sense of power.

Mark was waiting for him outside on the street. His chin was doubled against his chest, his hands were pulling his black curly hair into a pony tail. Mark tied a rubber band around his hair and flicked a few loose strands into place and lifted his head to disapprove a gang of giggling girls who were admiring his looks.

Charlie climbed in and lit up a reefer and played the radio. He decided upon some cool jazz for their soundtrack. Mark drove fast out of town. Charlie became silently philosophical. He considered old machinery and future wars.

— This plinkety plonk cocktail crap is getting to my head, said Mark, can't we find something else?

Charlie turned the dial, found an afternoon play acted by adults pretending to be children. They listened to it for a while, an occult suburban adventure interspersed with a sub-plot about a kiddies' piano competition that involved lots of cute crappy playing of Beethoven, and when they'd heard enough to be able to follow both the plots he switched over to jangling guitars, they grew tired of that, spin the dial — a saloon singer who'd lost his girl — a weather report — a financial phone-in — the cricket scores — the robust slash of a German orchestra. They got bored by the marching horns at the same time and Charlie tuned in to the Applause Station for the rest of the journey. They put on their fascist dictator faces and inclined their chins towards polite ripples of handclaps, to mild swirls of laughter; simultaneously they lifted proud hands into the appreciative air which became heavy with great waves of hollering cheers, wild ovations, cries of *Bravo!* and *More!* and

Author! Author! The car sped on down the motorway as Charlie and Mark, with fists raised, little shoulder bows, their faces proudly turning left and right, solemnly acknowledged the roars and shouts and clamours and yells, which steadied into a slow adulating rhythm of stamping feet and tuneless whistles.

4

CHARLIE DIDN'T LIKE to be alone. Sometimes, even, he got scared. Walking down sunlit streets with no one beside him. Driving his car along open avenues where the branches of trees struggled to hold back the sky. Waking up in the morning with only the faraway sound of a radio for company. Usually the combination of solitude and space would only gently disturb him, a murmur, a grumble, a hint of the end pressing upon him, no more, but sometimes — the chancy combination of light and solitude and doubt and, half-expected, unstoppable, up the fear would rise, pumping his heart, stifling his lungs, filling his head with the most disastrous possibilities. Charlie didn't like to admit that this was why he employed a driver. He tried to tell himself that he did it for the freedom to drink and smoke and dream.

Late one morning Charlie had discovered Mark about to break into his Scimitar in the company car park. A bunch of keys in his hand, trying each one in turn as his body twitched in time to the music he was listening to on his Walkman. His hair had been different then, a defiantly neat and old-fashioned short-back-and-sides. He wore tan overalls, LPC stencilled on the back.

Charlie had crept up behind him.

— Why, he had asked, are you breaking into my car?

To reach the boy's attention he'd lifted away the headphones and repeated his question. Mark had turned around slowly, didn't rush, didn't panic. Just turned around, grabbed the head-phones back, and sneered at Charlie.

— I weren't, he had said.

— I think you were.

— This car, Mark had said, rather more snottily than the occasion warranted, is a plastic heap of shit.

A pause. They had examined each other for a moment as if they were waiting for the other to draw the first knife.

— If I were going to nick a car — and I'm not saying I would, don't get any ideas — but if I were, you'd be out of your fucking mind if you thought I'd even *deign* to consider nicking yours. To be honest, I was testing my keys.

Charlie had laughed. He admired the boy's arrogance and the easy way he had with righteous anger.

— Testing your keys.

The boy resented Charlie's laughter. He looked at him even more suspiciously.

— They're new. I bought them off a bloke.

— Would you like a job?

— I have a job.

His job was driving a forklift truck around the factory yard, shifting drums of raw plastic feed from one corner to another. Charlie offered him the prestige post of becoming his personal driver, double salary, no uniform requirements, freedom to use the car whenever Charlie didn't need it, the willingness to be on twenty-four-hour standby. After a cold-faced internal struggle, a search for hidden catches, the reassurance that in no way was he being made fun of, Mark had taken the job. The arrangement had proved satisfactory for them both. They worked well together. When Charlie needed company he found a girl or called his driver. (The two did not mix well: Mark disapproved of many things and sex with strangers ranked high on his list of worldly evils.) Mark was always a telephone call away. Sometimes they ran into each other at Howard's house. Charlie there to keep an eye on things. And Mark to talk to Simon in that creepy culty way that left Charlie cold. Mark and Charlie saw more of each other than anybody else. Neither trusted the other. They were almost friends.

Mark dropped Charlie off in Paddington, and Charlie was soon naked in a candle-lit room. Belly wobbling, eyes shut, flesh rubbed with a mix of essences that comforted the body and found the soul through the avenues of his skin and nose. Inga's miraculous fingers massaged his body so delicately and hard. The massage parlour was called Scandinavia, where the cuties dressed in leopard skins and leathers and traded in their London names for Swedish ones. The rooms were called saunas, plastic walls made to look like pine. A laminated menu, headed Smörgasbord, gave the prices for the handjobs and blowjobs and full body massages, and the house speciality, Booblie-Wooblie, served up by a buxom Thai girl called Ingrid who bounced the customers' oiled faces between her oiled breasts. But when Charlie came here, on Wednesday evenings, all he wanted was Inga's aromatherapy. It relaxed him. It eased his spirit. It made him feel young.

A dream had brought Inga from Malmo to Paddington. It was the dream of owning her own Swedish massage and aromatherapy centre. Inga was the only Scandinavian girl in Scandinavia. She was the only masseuse in the place who did not fellate or flagellate or masturbate or administrate booblie-wooblie. She made only three concessions to her workplace. Under her open white lab coat she wore a leopard-print swimming suit tight over her sinewy body and sporty breasts. On her feet she wore a pair of white high-heel shoes. And, as she worked, carefully, considerately, concentrating upon each part of the body in turn with her oiled fingers, she gave with her own peculiar version of a porno girl's patter.

Inga was working on the backs of Charlie's lower legs. The rest of him was hidden under overlapping layers of towels. His eyes were closed. His face was peeping through the hole in Inga's warmed massage bench.

— Worried about you, big boy, she said, looking for and eventually finding the pulses in his knees. Too much bad food. Too much no exercise. Too much with the drinks and the drugs and the tobacco. You're killing yourself, lover man.

Charlie grunted his response. He wanted to drift. Her hands moved up his thighs and slapped on to his bottom. A few karate chops and he was feeling loose. Too little time had passed before the attention had moved to his lower back, but that had its advantages too.

— Yeah, right, that's good. That's nice.

Each vertebra in turn received Inga's special healing magic touch. Charlie sighed. Charlie felt young.

— Your heart. You not young man, stud. Check your heart. See doctor.

— Uhm. That's nice. I did once. Thought I'd had a heart attack. They said it was psychological. I said I didn't think I had a psychology. They gave me the bill and a diet sheet and told me to get the fuck out.

They'd run him through a battery of tests. It had cost him. Even, for no reason they would ever tell him about, they'd stuck a catheter up his dick. Clean bill of health. He was diagnosed as suffering from panic attacks. His world had changed after that. He started to suffer panic attacks. He hadn't driven a car since then, or taken any other means of transport by himself. Airplanes and tube trains and solitude especially made him feel like he was due to die.

— When? Five years ago? Things change, lover. Hey. Favour. Yes? Cut down on the bad things. Tell your baby yes.

She was at his shoulders now, unknotting and untying the stress.

— Sure thing, baby. Yes yes yes.

— Take responsibility for your body. Turn over now.

— Hyeugh, hyeugh. I take responsibility for *this*!

His dick was suddenly erect and proud and daring her for a favour. And just as suddenly it was flaccid again, courtesy of a special pinch nerve-hold she'd learned back in Sweden.

— I *hate* it when you do that. And why do you have to talk like a fucking Siamese whore anyway. And what's that oil you're using?

She covered his groin with the heaviest towel and unravelled the one on his chest. She rubbed her hands briskly together and got to work with the flesh.

— Grapefruit and basil oils mixed together, big boy.

— Sounds like a fancy fucking salad.

And then it was back home to his pad in Battersea, by the river. Bachelor night. On Wednesdays Charlie, his soul strengthened by Inga's aromatherapy, always tried to spend the night alone. He listened to jazz records and watched the river. He inspected the bedrooms on the opposite bank through his binoculars. He followed the progress of speeding cars and the manoeuvres of the traffic police. He kept his mind empty of business and girls and fear. He didn't smoke cigars on Wednesday nights. He wanted the scents of elemental oils to linger about him, pure.

The rest of the week was always busy. The week was a series of slots that Charlie slid himself into. Accident, chance, loneliness, they were panic's friends. Thursday night he met Rocky's girl. Friday night, poker night. Saturday, it was financial affairs in the morning, football in the afternoon (cricket, a poor substitute, in the summer), a restaurant at night. Sunday he played tennis and drank and always went to the cinema, often twice, sometimes with a girl or else with Mark. Monday and Tuesday at the Heel Bar, with Tuesday always finishing off at The Merry Mogul Indian restaurant. But Wednesday night was Charlie's special night, when he allowed himself solitude and thoughtlessness.

5

THURSDAY LUNCHTIME, AND Charlie was somewhere he liked. He lit a cigar. He propped the photograph up between the pepper and salt pots, it fell flat and he left it where it fell. He waited for Chatterjee to arrive.

He was glad to be here. He ate lunch at the Happy Eater every Thursday, Chatterjee or no Chatterjee. It invariably made him feel good. Everything was plastic. They sold Family Novelties in the fake-brickwork corner, brightly coloured things that looked like dildoes in cellophane wrappings. In the open kitchen dark immigrants in green aprons and white paper hats slapped pineapple rings down on top of pink slices of gammon ham. Red plastic tables and green plastic chairs. Plants everywhere, and all of them plastic, polyester silk, in the window bays, hanging down in baskets from the ceiling, pinned to the wooden-effect walls above the kitchen counter. Bright yellow radiators co-ordinated with the yellow trim at the base of the partition walls. Laminated menus, beige for the Main Events, blue for the Happy Endings afters.

— Would you like to order now or are you still waiting for your party?

Hello waitress. Stiff-faced cutie dressed like an air hostess. This was his favourite part. Charlie always ordered randomly, his little flirt with chance, no cheating by looking at the menus first, he just called out numbers from nought to a hundred in the joy of building his own surprise.

— Might as well get the ball rolling. I'll start with a seventy-seven please.

— A Strawberry Fields?

— Uhm. Yes please. Thank you. And a, a thirty-eight.

126

— Howdie Doodie? What sort of toast would you like with that?

— Good question. White I think.

— Would sir like tea or coffee? It's included.

— Both. And a beer.

— Lager or bitter or stout?

— Lager I suppose. Oh, and give me a number four as well.

— That's the Chicken Goujons, served with a delicious and tangy minty-yoghurt-sauce.

It was risky, the last one, he knew it at the time. She'd already been ominously quick to put the names to his numbers. He'd gambled that she wasn't one of those waitresses who knew the low-number dishes off by heart. But still, even if he was asking for it, she should have known that this was his game, not hers. It was revolting the way she said the description so proudly as if he'd be pleased for her.

— Whatever. And could you bring it all at the same time? Thank you (he peered at her bright yellow name badge, chirpily bouncing between small pointed breasts), Tina.

Outside, through yellow-rimmed windows, a red children's slide with the lonely face of a cartoon dinosaur, a clump of grass, and the customary traffic jam on Western Avenue. Inside, under yellow globe lamps, happy families and solitary salesmen; his nearest neighbours were a child playing stickily about with his Crazee Nuts Sundae and on the other side an emaciated man in leathers telling the waitress about good friends who had died horrible deaths in motorcycle accidents. It was a carnival of plastic in here and out there riches awaited so why did a mystery photograph and a waitress with a good memory make him feel as if he wasn't on top of things?

Mr Chatterjee scurried in at the same time as the food arrived. Mr Chatterjee didn't want any food or any tea or any coffee. Mr Chatterjee was here for business. Charlie didn't want to think about that yet. He read to Mr Chatterjee the food descriptions from the menu as Tina unloaded his choices from the tray. Strawberry Fields was A Haze Of Strawberry Yoghurt Ice Cream Layered With Strawberries And Meringue Under A Mountain Of Fresh Cream And Strawberry Sauce. Howdie Doodie! (the exclamation mark pleased him enormously) was

127

Rashers Of Best Back Bacon Served With Fried Free Range Egg Accompanied By Grilled Fresh Tomato, Mega Hash Browns And Your Choice Of Toast And Coffee Or Tea. Tina had already spoiled the surprise of the Chicken Goujons and she would suffer for that later with a tip as mean and pointed as her tits.

— God I love this place. You sure you won't dig in? There's enough for two.

It took more than taste sensations or plastic aesthetics to press Mr Chatterjee's buttons. He was a small man, this Chatterjee, had some children, originated from Asia (Charlie never could remember from where — India? Pakistan? Bangladesh? Sri Lanka?); he was rounded at the edges, balding at the top, hairy around the mouth, and he sweated inside an unseasonal woollen suit. He spoke very nicely, like an actor, and he always began, courteously, with small talk.

— I have lunched already, thank you. Forgive me if I have kept you waiting. The traffic has been quite terrible. The weather, though, is pleasant, and more than makes up for it, don't you find?

— Try some of my goujons. Just to taste.

— You are very kind, but again I must to say no. I think, if you do not mind me saying so, that it would be advisable to conclude this as peremptorily as possible. It is not in our best interests to be spotted here.

— We shan't be. Do you know anyone who comes to a place like this? I don't. That's part of its charm.

— None the less. We are close to your place of work. There is a petrol station beside us. We cannot say with any certitude that there is no chain of circumstances that will not lead someone you or I know to driving — excuse me — to be driving slowly beside that window. And then . . .

— Fair enough, said Charlie. He hesitated for a moment between his Mega Hash Browns and the sludgy ruin of his Strawberry Fields. Charlie discovered a long time ago that whenever he was involved in negotiations where bullying was not suitable then it paid to hide his thoughts beneath a layer of greed or stupidity or both.

— Mr Levy. Please. You will excuse me if I insist we apply

ourselves to matters of business. Have you discovered any impediments to the plan?

— None.

— Do you retain your earlier enthusiasm for it?

— Entirely. All systems go.

The first answer was true. The second was partly false. Charlie did retain his enthusiasm but it had been joined by a feeling of dread and he swung between the two like someone who had gone to watch the circus and one moment he had been down in the crowd cheering a near naked girl prancing about on a horse and the next he was suddenly swinging through the air on a trapeze: and of course it was exhilarating being higher than anybody else in the world, and safety was only a step away, but no matter how much he lunged with his feet he could never regain that empty platform at the side, and why was there a pain in his chest and was that a *screw* pulling loose from the mounting at the top?

— So when, Mr Levy, will you be in a position to sell?

— A year? Two years? Tomorrow?

— I beg your pardon for pushing you, but which?

Charlie dabbed the corners of his mouth. He contemplated the plates in front of him. He penetrated the Howdie Doodie! free range egg with a quarter of toast. Yolk oozed satisfactorily out. He smeared the toast into the egg and munched it down. He wished his appetite were stronger.

— I found this photograph the other day. It's my father and his brother. They started the company. My father's the strong one.

Mr Chatterjee recognised that he had to treat this subject with respect. He turned the photograph with a careful fingertip at each top corner. He examined the faces. He nodded.

— He wasn't the brightest man in the world. My uncle had the brains and my father had the muscle. For a while he ran the business on his own and got into trouble doing it. Had to bring in his sister to keep a hand on things.

Mr Chatterjee reluctantly nodded again. Charlie watched him battle down the impulse to sigh.

— Listen to this. This'll give you a handle on him. It was in the early 'fifties, when Tupperware was about to get big.

Tupperware was invented, I'm sure you know, by a Yank, Earl Tupper. My father had the idea to get a licence to manufacture it over here, give him some credit, he spotted the potential. But this is the thing, he was convinced, nothing could persuade him otherwise, that the Yank was an aristocrat. That Earl was his *title* for fuck's sake. Insisted on writing letters addressed to the Earl of Tupper, telephone conversations when he'd say in his best courtly East End manner that he wanted to speak to *Lord* fucking Tupper. Unbelievable. Not surprisingly, they gave him short shrift. It was after that that he got Freda on board.

It's an old family story. Charlie had never believed it but he'd always liked it. It was Dinah, he thought, who told it to him first, making tipsy mischief at his first wedding party.

Mr Chatterjee allowed himself to sigh. Charlie couldn't stretch this out much longer. This was their third meeting. The first took place as if by chance just after Howard ran away. Charlie was at a Wednesday football match, Queen's Park Rangers, Loftus Road, the last home game of a surprisingly good year. The referee blew his whistle, game over, a lap of honour, see you next season, thanks for everything, and as the players finally left the pitch, milking the applause with waves and kisses, Charlie had stood up, shared a final joke with his season ticket neighbours, joined the line of contented fans waiting to get away to summer. And the little Asian man had been there beside him in the aisle, whispering so discreetly about the weather that it had taken a while for Charlie to realise that the man was talking to him. He wanted a meeting with him, Mr Chatterjee had said, to speak of something secret that would be to his advantage. Charlie thought he was about to be offered drugs or sex or something stolen and unwanted but then Mr Chatterjee had persuaded him of his serious intent by citing serious things — the address of the factory, the registration number of his car, the recent change in his position at the company — and Charlie had decided then that the man was an accountant who worked for the Inland Revenue and operated a blackmail insurance scheme on the side. He had heard of these renegade accountants. They needed to meet at a convenient secret place at the earliest opportunity and as the next day was a Thursday the Happy Eater down the road from LPC was the most secret place that Charlie had been able to come up with.

They met there. And the sinister money man had turned into a benevolent property man with a secret scheme, possibly criminal, that filled Charlie's heart with joy and his imagination with a sense of the infinite. They had talked and they had negotiated and Charlie had named a figure from out of his dreams and Chatterjee had agreed to it. And now it was about to be confirmed, no paperwork necessary, just a Happy Eater accord between two serious men who had discovered a way to bend the world to the shape of their own will.

— I can't give you an exact date, Chatters, this is a dynamic situation.

— My principals need a deadline, Mr Levy. I do apologise for insisting.

Find Howard, clinch shares, clinch deal. This could, as he said, happen tomorrow. On the other hand...?

— Mr Levy? Just to be on the safest side we might choose the maximum time available. Why don't we say two years?

— Why don't we? Two years.

There. It was done. A deadline far away in the shapely future. Unthinkable that it would take as long as that. Now, gently, Charlie had to introduce the topic of a reward for bringing it home earlier. Thoughtful Chatterjee was keeping abreast.

— If it could take place sooner that would bring, I think I can promise you, great *pleasure* to my principals. But you do understand of course that there can be no question of any extension.

— Goes without saying. And we are still talking the same basic figure?

(Charlie coughed, looked around, lowered his voice to a tender, almost priestly whisper.) Three million pounds. To me that is. Ten per cent of that to you?

— Of course. I am not a greedy man. I only wish to accumulate enough capital to help my younger son to a nice location for his practice.

— Absolutely. Young Doctor Chatterjee of Harley Street, anything less would be unworthy. And your other children? (Charlie hunted the messy surface of his memory for their names

131

and found only unlikely shadows — Fatwa? And-And?) How are they?

— Adnan is a fully qualified property surveyor. Tara dances the ballet.

— Very noble professions. Must come in handy too in your line of work, the property surveyor I mean. You sure you don't want a taste? These Mega Hash Browns are terrific.

— Quite sure. I think it would be best for me to make my exit now. We should meet again in exactly a month's time. Communicate upon our progress.

— Let's.

— It has been, as always, Mister Levy, a pleasure.

— Quite. Likewise.

A final handshake, an exchange of nods communicating mutually suspicious understanding, and off went Mr Chatterjee, brisk little walker, attaché case carrier. Charlie pushed away the food, settled his bill with plastic, and left a generous cash tip for naughty Tina despite the damage she had done. He put his jacket back on and walked up Western Avenue to his factory. Panic reared up at him and Charlie stared it down. Charlie could still walk alone.

A man steps into your life from out of the blue, a dapper, sweaty little man with secret knowledge that he knows how to put to profitable use. Don't look a gift horse in the mouth. Look before you leap. Every cliché had its enemy cousin. And let us remember, above all, the lesson of Lenny Leverton. Charlie had made enquiries about Mr Chatterjee, careful, tactful enquiries. He recorded their Happy Eater meetings on a micro-cassette recorder. Mr Chatterjee was on the level and, just in case he wasn't, Charlie had protection.

Facing away from Western Avenue, looking so old and stupid, an obsolete prison, the LPC building. Only one window at this side, low on the ground floor, where the workers had their Recreation room, which Charlie visited once a year in the interest of staff relations for the annual table-tennis competition. Charlie walked along the lane, entered the half-empty car park. He looked at the factory and he saw beyond it. With Mr Chatterjee's help he had stepped out of history.

It had been there too long, LPC, dripping plastics into the ground, sending out its loads of mock tortoise-shell combs and brushes, buttons, cases for battery razors, computer casings, toys for children to stick on the ends of their pencils. It was gone — just imagine it and it was: the building smashed by the swinging, crane-mounted lead ball of Charlie's ambition, levelled. The empty grassland behind, where the company team plays swampy football games, was already, if he just almost shut his eyes, filled with a gleaming arcade of shops. Was there a stream? Where ten-year-old Lawrence in grey school shorts, holding a fishing net, his plimsolls and socks left on the shallow bank, once ripped apart his ten-year-old shins? There was no stream. Only a fountain, two storeys high, sparkling with lights, for the indoor shoppers to gather around. His father was dead. The future was here. The water had been drained, liquid concrete poured, it was all ready, the whole site, ripe: LPC was gone and in its place an access road to the car park of the shopping centre or palace or petrol station or mall, it hardly mattered, they could build what the fuck they wanted; and the arthritic ghost of LPC would sit out the rest of the dance, and when there was no one left to remember its name it would glide into a stain on a new plastic wall and shrink and die.

The thing had run its course. It was time for the strong man to lift up the family blessing and throw it aside. So why then did Charlie suddenly look over his beefy shoulder and feel as if there was someone older and better than him watching?

6

STRIP LIGHTS, FITTED carpet, framed product photographs on the walls, an empty, burnt-bottom coffee-maker in the alcove behind the horseshoe sofa, large pretty girl behind computer terminal and desk, and, on a corner of the desk, completing the ideal office picture, a cocky boy in a striped shirt lounged.

— Nice lunch?

Charlie Levy, Managing Director, Company Secretary (acting), Chairman of the Board, fortune's anxious darling, let the wooden office partition swing behind him.

— Thank you Babs yes, how's the . . . er?

— Much better thank you, you had some calls, they're on your desk. And there's a Development meeting at three, and a Feel Good meeting at four.

— Arnie reported yet?

— Oh right. The weird thing. Not yet. I'll get on to him.

The weird thing was that a couple of nights earlier two machines had gone rogue, just started up of their own accord, spewing out goblins and combs, a giddy party of unprogrammed production.

— Tell him I want to talk to him today . . . Busy are we, Gary?

A spaghetti western exchange of raised eyebrows and sinister half-smiles until, before he was about to lose his other battle, with a Howdie Doodie! fart, Charlie pushed on, up the staircase, along the executive corridor — neat facing rows of department doors, behind which laziness and fanaticism and greed were doing their things — to the sanctuary of his office.

Charlie sat at his desk made of moulded plastic. He chopped up two lines of cocaine with a credit card, sucked them into his nose with a milk shake straw. The cocaine thrilled his head and panicked his heart. Charlie shut his eyes. Charlie pounded the desk with his fists. Charlie said *Yes!* and kept saying it until the veins in his eyelids stopped throbbing. He spun in his chair. He got up. He admired the things on the walls, including, pride of place, above the mantelpiece, between tiny industry trophies, a white button mounted on cream card and framed in imitation silver. A chrome-effect venetian blind covered the window that otherwise would offer the dismal view of the wasteground. No longer dismal, happy view now, infinite potential: Charlie rolled up the blind, the full glory of the wasteland.

He had to find Howard. Sneak away some of his shares. Before he could own seventy-six per cent of the company he had to make sure that no one else could own twenty-five. And he had still to persuade Dinah to part with her shares. He needed more money. The campaign was expensive. Shares, even in a worthless company, cost. Charlie took to his desk again, checked his messages. One from his broker, the Classic Club fund was growing. Another from the smart boy he'd commissioned to find Howard. The smart boy was still not proving himself to be very smart. Or maybe he was, getting rich on Charlie's money and not giving anything back in return. Charlie crumpled the messages into a ball, tossed it into the air and flicked it with the outside of his shoe into the middle of the wastepaper bin. He pushed the pile of papers for his attention to one side and picked up the telephone and dialled Howard's number.

— Vivien, hi. Charlie. Any news?

There was no news. The last anyone had heard of missing Howard was when Katz had made the money arrangements. Since then nothing. People didn't just disappear. It wasn't possible. People get killed and an elderly gardener digs up the body and his fluttering wife calls the police. People get amnesia and turn up in the hospital with a bandage around their head. People hide and a smart boy with the latest technology finds them. Charlie invited Vivien out for a drink. Vivien said she was going to be too busy. Charlie blew her a kiss.

— Ciao baby. Call me if you need anything.

Everything had been set up, everything had been almost ready, the machine for profit that Charlie had so carefully devised. Howard was being eased out of power, and then Charlie or his agent would have quietly bought his shares and the machine would have gone into production. And then Howard in his ineptitude had run away.

Charlie made more telephone calls, with purpose. The smart boy to have somebody to shout at. He did. His second childless ex-wife to find out if he could charm her into lowering her expense requirements. He could not. Rocky Sporgente's girl-friend to confirm their date for the end of the day and to be thrilled by the flatness in her voice. His driver to tell him when to pick up Rocky Sporgente's girlfriend. Charlie tried to get his driver to confirm the information but Mark broke the connection in a futile show of rebellion. This annoyed Charlie. Mark should have more respect. He was like a guru to the boy.

Charlie had never had a guru. In 1968 a friend called Kevin had gone to Poona because he had been told his true spiritual self would be found there. Kevin and forty-two others spent three days in ecstatic communion with the swami. They danced, they sang, they daubed red blobs onto their foreheads, they went on bicycle rides along the banks of the holy river. At the end of those three days, in a sun-blessed, tree-lined courtyard, where the scent of saintly sweat mixed with lotus flowers and incense, the swami dispensed to his forty-three acolytes their true spiritual names. Nigel became Anand. Martin became Sri-kanth. Deborah became Meera. Gilly became Bapsi. And at the end of the line Kevin became Kevin.

It might have been the exhausted failure of divine inspiration, it might have been a sly guruish joke. Or it might have been the delightful cosmic truth. Kevin took it as the delightful cosmic truth. He returned to London dressed in saffron robes, his Brian Jones page-boy cut gone, just a blond hair horn poking out from his shaved head. And even when he became a sciento-logist four years later he stayed true to the holy name of Kevin.

That for Charlie had ruled out religion as a way to replace a dead or missing father. Charlie had joined the company in 1973. He'd fought against the idea at first, envied his brother Maurice who had cashed in his company chips on his twenty-first birth-day and gone to hunt for drugs and pussy out in the world where, presumably, he still was. But there was something inside Charlie, something which he thought of as timid and soft, and occasionally as righteous, which had made him do the right thing.

A grumpy salesman called Sid, who suffered from a catarrhal problem and a beautiful second wife, was supposed to take Charlie under his wing. Sid had been at LPC almost from the beginning, when it was still called the Levy Plastics Company, before it became the London Plastics Company and then, demurely, LPC. Sid was to instruct Charlie in the ways of the order book. Together they were meant to visit customers, to sell Universal Unbreakable Combs and mock-tortoise-shell brushes and buttons and hair clips and hair rings and models of Second World War soldiers and American rocket ships. They would attend the toy fair in Harrogate, the housewares fair in Blackpool, the hair care fair in Scarborough. *It will be your education*, Freda told him. *Sid will be your Oxford and your Cambridge both. If your Father had still been alive, God rest him, or your Uncle* (a puckering around her mouth that suggested she was about to spit) *then they would know what to do with you. You would get your hands dirty, you would be taught machines. But instead, in their place, I must try to do what I think is best.*

Charlie and Sid reached an arrangement. Sid would introduce him to the buyers from the big chains. Charlie would use his own initiative and the company money to influence things in the tryout shops. And for the rest of it Charlie wouldn't bother Sid and Sid wouldn't bother Charlie. Together most mornings they would set off, in Sid's dusty Rover, the boot filled with samples, and when they had got around the corner Sid would stop the car, Charlie would get out, neither would look at the other, and Sid would sluice some phlegm around his throat and say — fixing the car back into first gear, impatient foot poised over the accelerator pedal — *if you're not here at five there will be trouble*. Charlie was seldom there at five and there never was any trouble. Because Charlie had taken the precaution of following Sid on two separate days and snapping some polaroid photographs of the Chief Salesman (Products) in the company of two of the doggy blondes he liked to spend his time with.

The Chief Salesman (Machinery) was Howard's sponsor. A thin bachelor called Joe, known as Joe the Poet, who had suffered an unmentionable tragedy in the War and was writing a blank verse cycle that appeared in half-yearly instalments in an unfashionable literary magazine, which turned out later, a scandal that Howard once bored Charlie with on a plane ride to an engineering fair in Stuttgart, to be financed entirely by the CIA. Joe the Poet looked after Howard in a way that Charlie at the time sneered at and sometimes coveted.

The telephone buzzed, the past dissolved (ever since Howard disappeared he had been getting maudlin, sometimes, even, nostalgic), dutiful Barbara called him to the Development meeting. Charlie walked, like a gangster, out of the office, down the corridor, into the boardroom where all the furnishings — this was Charlie's first initiative as the man in charge — were plastic.

Charlie settled. He helped himself to a chocolate biscuit from the nearest silveraria plate and filled a polythene tumbler with water from a polycarbonate jug. Around him, the war zone. Different departments competing for slices of the budget. Shiny boys in shiny suits who wanted all the money to go on special promotions and lunches. Keen weasel boys eager to show off their latest crazy plans. Cosmeticised girls with insincere lips that promised fellatio. These departments hated each other. All they shared was company notepaper and the desire to murder one another. Instead though they'd killed Mrs Buller. The weasel boys were making their pitch. A man's suit that could shine different colours and never needed to be ironed. A self-cleaning car chassis. Back to the roots with a hairdressing kit. A way to manufacture computer casings that had no rounded edges. Eddie wanted to diversify into metals. Say no to that, straight off. William had a thought about costume jewellery and Bill called him a poof. Ron said that PVC raincoats were making a comeback. Arnie from America hadn't been able to figure out why those two machines had set themselves off but he did have a plan that nobody could follow that had something to do with superconductors.

At the centre was Charlie, chewing on a cigar, hooding the boredom in his eyes. Opposite him were the twin paintings of

138

Sam and Danny, the hard man and the visionary. Side by side portraits of the founders of the family fortunes in curlicued gilt frames each with a little bronzoid lamp fixed to the top to shine a halo over the dead heroes' dark foreign heads. Next to them, little Freda, in a man's business suit, standing at an elephant of a desk piled high with buttons and combs, with spectacles in one hand and a blueprint rolled up in the other, her portrait respectfully a little lower than her brothers. There was a small vomity splash of a thing beside her which represented the brief reign of Howard. Charlie decided it was time to commission a portrait of himself. Try something in acrylic, or maybe a sculpture. Valerie in the office had artistic tendencies. He would get her on the case.

Charlie lighted his cigar. Whenever there was a pause people looked to him and Charlie didn't say very much because if you said too much you gave yourself away. (Hard to keep track of who he had used and who had tried to use him and who owed who favours and who he needed — did he need any of them?) A simple yes to Ron, a wink to Arnie, shakes of the head to Eddie, to William, to the rat pack boys Angus and Michael and Phil. William poked Bill in the arm, Bill made as if to punch William in the face, a cup of coffee spilled, steady Eddie stepped in between. All the fun of the fair.

The Development meeting slid into the Feel Good meeting. In came Leslie, looking shifty and sleepy and sad. Had to be careful with him, needed his shares and the ones he had from Freda, maybe wait a while before revealing that his sly little approaches to Saul Cohen's daughter had not gone unnoticed. Guido from the shop floor, thought we'd sacked him years ago. Gary, the marketing manager, sleek boy with a very round skull and crispy lacquered hair. He played the electric piano every year at the Christmas party. Barbara arrived with Valerie, they found unimportant seats near the end of the table. The rest of them came in, loping and striding and sidling and marching, the lawyers and accountants and salesmen and machinists, Accounts over there next to Financial Affairs, Public Affairs jowl by jowl with Direct Mail, they all took their places, tried to look as if they were ready to feel good.

— Yes! shouted Charlie.

— Yes! came the echo, mixture of shouting and murmuring, not all together; some were too shy to do more than whisper, like guilty Leslie who, in all the family celebrations he has ever seen him at, had never sung happy birthday, just sometimes, if he thought he was being watched, would mouth the words bashfully.

— Yes!! shouted Charlie

— Yes!! they shouted, louder this time, more in unison.

— We can do better than that. Yes!!!

— YES!!

— That's good. Now come on. Not everybody's joining in. Leslie. You.

Silence. Everyone looked at Leslie malevolently, glad it was him this time to have the public sword pointed his way. Leslie was just sitting there, head bowed, picking at the eczema on his hands, hoping against hope that there was someone else called Leslie in the room.

— Leslie.

Leslie shivered, looked reluctantly up.

— Me?

— Let's hear from you solo. Give it some throat. Yes!!

Leslie gulped. Blinked. Lowered his spotted balding head. He opened his mouth and a tiny affirmative peeped out.

— Yes.

Poor man, he was blushing. Shouldn't be doing this. Need to soft-soap Leslie. Fuck him. He'd been sneaking. He needed to relearn fear. Charlie knew no pity.

— Louder! Yes!!!

— Yes!

— Again!

— YES.

— Better. Now everybody. A big one. YES!

— YES!!

— YES!!!!

— *YES!!!!!!*

Charlie then started to clap and everyone joined in, sustained, rolling applause, thunder-clapping, wolf whistles, some hollers and shrieks. Angus from Chem-Dev trilled in with a yodel.

— All right, settle down, that's great. It's love your neighbour time. Gary, you start.

Charlie sat back. He'd done his bit. Gary from Marketing praised the latest brochure from Nikki Public Affairs. John Direct Mail said he liked the nurturing work of Sally After-Sales. Say something good about somebody else and you'll probably hear something good about yourself. Gary and Nikki, an unlikely political alliance, perhaps they might be . . .? Charlie drifted off while allies said nice things about allies, and enemy departments silently built their trenches.

Not so silent. In the corner by the window William and Bill were facing up to each other. Taking it in turns to poke each other in the chest and jerk their heads and say *Yeah?* The company solicitor Katz busied himself squaring off sheets of paper with his suspiciously delicate hands. After-Sales was criticising Sales and jabbing a finger in the direction of his water glass. Distribution was blatantly laughing at After-Sales. Direct Mail was talking quietly now into a mobile telephone. Public Affairs nodded agreeably at each jab of After-Sales's finger. Accounts and Shop Floor were playing battleships on their handkerchiefs.

Somewhere a wrong turn had been made. Charlie had made mistakes in his term as Managing Director, so had Howard. The people who worked for them didn't like each other. The company manufactured too few products and bought in too many and until Charlie had taken charge there had been nothing spent on research and development. Profits were diminishing. Installation of the computer system had hammered them silly. Something needed to be done but nobody seemed to know what it was. My destiny, Howard had said at one of these meetings, was meant to be quieter than this.

Charlie wondered if he ought to fire everybody. Begin the thing again. Take Uncle Danny's machines and patents and find them a more modest home. The machines deserved better than this. Plastic deserved better than this. Those sleek boys in their Italian clothes were concerned only with their departmental budgets and prestige. Everything had succumbed to inertia. Something had been forgotten and Charlie had forgotten it too. Michael jumped in between William and Bill. Angus and Phil

141

pulled them apart. Someone was spitting. Nikki, the ungrateful bitch (worst mistake he ever made was not to sack her when they were through), was giving him one of her you-better-promote-me-or-else looks.

— Perhaps, said Leslie, daring to speak, we should open a window.

Everybody nodded gravely. Sales ran over to the window and lifted it open. A sudden warm breeze forced itself in, scattering the solicitor's papers. Water rippled in water glasses. Chocolate biscuits were melting onto their silveraria plates. Charlie gave a big broad grin and tapped his glass with his cigar clippers. The department heads slowly fell silent. Charlie stood up. He emptied a glass of water into his throat and held the glass out for Shop Floor to fill it up again. He looked around the room. He creased his face into what he hoped was a gung-ho grin.

— Well done, he said. Well done.

This was a good start.

— I mean that, to each and every one of you. You all deserve a bloody great pat on the back. Times have not been easy of late. We've been up against it. I say that frankly. Companies like ours are not being given any breaks. I've been listening very carefully to everything that's been said here today and I want you all to know that I pay particular attention to all your opinions and thoughts. I'm glad we have the confidence and the trust in each other to speak so candidly. Profits have been down and to some of you that signifies failure. Think again. We're still making things and we're still selling the things that we make, that's the main thing. I remember something my father, Sam, used to say . . .

He told this kind of story every time and he was doubtful if anyone still took the trouble to listen. The words pattered out of his mouth and he had long ago given up wondering if they made any impact. But in this moment, for this moment at least, his own heart was in it. The grand historical line. Danny to Sam to Freda to Howard to Charlie. They were all listening to him, to his corny story about captains and ships and hands on helms, even William and Bill had stopped making monkey spitting sounds at each other. The hot breeze had diminished. The sunlight made rewarding patterns in the water glasses, tentative rainbows arced across the table. He noticed, for the very first time, that Leslie had very nice teeth. For this moment, as long as he kept on talking, things made sense again.

— . . . and I want to thank you Leslie, and Gary and Edward and Guido and Nikki and Sally and John and Valerie and Barbara and, and Jennifer, and all the Dev boys, Chem and Elec and Comp and Bio, and Chris and Arnie and Mister Katz and I want to thank . . .

Charlie thanked a lot of people and everyone was nodding and some were smiling.

— . . . and I want to remind you of what we are here to do . . .

He reminded them of the solemnity of the commercial mission and the grandeur of manufacturing and the joys of team work. And when he was done the feeling endured, and he drank his water and everyone clapped, even William and Bill, and Charlie had the hope that perhaps everything was going to work out, true to the dead shades of his father and his uncle, and who the fuck cared about the solitary future or the death of Mrs Buller, and Chatterjee could go take his clever plans somewhere else, and so what if Howard was missing, he would find him eventually and they would cleave to one another like brothers, and he felt enthusiastic about everything again, he had convinced the people who worked for him and therefore he had convinced himself, and the skirmishes and the squabblings were all things of the past.

— Anything else? No. Then let's go back to work. Let's kick some fucking arse, and let's give ourselves a bloody good slap on the back while we're doing it. All right?

A final yessing session, roaring out confidence and pride and animal spirits, and then everyone meandered away, to offices and machines and cars. Charlie walked along the corridor, his arm around Leslie's tight shoulders, telling him a dirty joke about an elephant and a pygmy that suggested in the way of its telling that the two of them shared an exciting, manly world.

7

CHARLIE REGARDED HIMSELF in the washroom mirror and splashed water over his face. He was making himself beautiful for her. Charlie smoothed down the creases in his forehead with the pressing iron of his hand. He twinkled a seductive smile at the cocky bastard looking out as he combed back his hair. He washed away the few ginger-grey clumps that stuck to the teeth of the comb. He opened the mirrored cabinet door and took out his emergency tub of hair gel. As he levered off the lid there was a moment, just as his fingers were about to dig into the creamy orange jelly, when he was half-puzzled by the finger marks already there, alien indentations that half-connected to the mystery of the photograph and the mystery of the two rogue machines that had started in the night of their own berserk accord.

If his own fingers would have stopped at that moment, he might have paid more attention to the question forming in his mind, and maybe it would have occurred to him that there was less cocaine than there should have been in his desk drawer, and that his personal copy of the Buller Report had gone astray, and maybe he would have found an answer, too, about those intruder finger marks, not his, longer narrower fingers than his — but his fingers dug in and that moment was gone and he fetched up eight little tip-wads of grease and he rubbed his hands together and he combed the grease through his hair and he flattened the hair along the sides and the back and plucked his pompadour back into its perfect shape. The cabinet door swung shut and he whispered to the demon lover looking out, *You're gorgeous, up and at 'em killer,* and all that was left of his half-puzzlement was a sense of disquiet and an empty

144

feeling in his belly as if it was anticipating something dangerous around the next corner.

8

CHARLIE MET ROCKY Sporgente's girlfriend at the usual place and the usual time. Even so, an imp of panic stabbed needles into his heart and lungs and legs. He sat in the back of the taxicab, worrying if the time when he could travel by himself was over, and he rubbed his aching heart. Perhaps Inga was right. Too much junk food. Too much coke. The pain only started to leave him as he walked down the stairs to the Heel Bar club where Rocky Sporgente's girlfriend would be waiting.

Rocky's girl was acting bored. She was always bored. She was beautiful and pale and blonde and she was the classiest girl Charlie had ever thrown money at. She always wore expensive loose clothes and had tastes for American whiskey and Pakistani heroin, and she spent a lot of time bitching about her boyfriend, Rocky, a stylish light-middleweight with some very sharp moves and a chin made of glass, who held a peculiar notion about the mystical nature of sperm.

— He's Italian, she had told Charlie at their first meeting, and Italian men are fucked-up anyway, mother-Pope-buggery, and at the beginning I didn't think there was anything especially odd about him. What's your name? You can buy me a drink. Bourbon.

Rocky was going to marry his girl when he became light-middleweight champion of the world. Rocky was a very popular opponent. He gave everybody a good fight, other fighters' reputations or brutality didn't scare him, he was fast and he was strong and he looked very fancy in the ring and, ahead on points, usually in the ninth or tenth round, he'd get caught with a good one to the chin and go down, unconscious, with a baffled expression on his handsome, unscarred face.

146

— And that's OK, I don't care that he's a loser, but it's this thing he's got about his sperm. He thinks it's magical.

Rocky wouldn't fuck his girlfriend. He wouldn't fuck her in the two months before a fight and he wouldn't fuck her in the two weeks after a fight. Rocky fought, on average, every two and a half months, all around the world, he was in demand, he made a lot of money to make other people look good; and all of this meant that every two and a half months there was usually a two-day period when Rocky would consent to let go of his sperm.

— And then it's crazy. You should see it. Builds and builds and of course he comes really quickly because he won't even masturbate, his sperm is so precious, and it's like he's shaking and screaming, eyes rolling into the back of his head, cursing or praying in Italian. And he won't come inside me, pulls out, likes to come, can you believe this? onto his *boxing gloves*. So I have to hold them over my tits (and meanwhile I'm doing the corniest impression of an orgasm which wouldn't fool anyone over the age of fourteen unless he's an Italian). And the whole thing's over for another two months and the sad thing is he really believes he's going to be world champion some day. I'd like another drink now please.

Charlie and Rocky Sporgente's girlfriend met every Thursday at the Heel Bar, and they usually ended up fucking at Charlie's place, but despite her jeers at him and her liking for pain, and despite the fact that she would never allow him into her own apartment (that would be a betrayal of Rocky, she said, who paid the rent out of his losings), and despite the fact that they usually got stoned together on uncomplementary drugs — he smoking marijuana, she smoking heroin — Charlie had found that he could talk more freely with her than he could with any person since the early days of his marriage to his first wife.

— I'm on to something, said Charlie.

— I think, said Rocky Sporgente's girlfriend, I should get a job.

— Come work for me. I'm frying big fish. I'm skinning cats by the bucket-load. Or better still, let's get married.

— Rocky would kill us both.

— He'd have to find us first. I'm on to something big and when it happens I'm off. I don't know where yet.

He hadn't said it out loud before. He wished he knew whether he was telling the truth.

— I'm serious. I want a job.

— You don't want a job. No one wants a job.

— I'm getting old. I'll be twenty-seven this year. I'm sick of being given money by men.

— Fuck that.

— I'm getting lines on my face. What happens in ten years' time? I'll be a drunk old hag talking to my cats and men on the radio. If I had a job I'd have something to fall back on.

— Forget about work. I tell you, I'm going to be rich, big rich, not Rocky small-time rich. And then we can do what we like. We can start a circus in Portugal or go sailing around Cape Horn.

— I've read a lot of dictionaries, did you know that? I'd like to become an etymologist or a philologist. I bet you don't know what the difference is. Get me another drink please.

Charlie got her another drink. Negotiated the difficult walk from their table to the bar and back between a crowd of closed-eyed actors gigglingly falling into each others' arms. It wasn't just actors who used this bar, it was gangsters too and fashion types and one or two footballers.

— Have you ever needed to find somebody? Have you ever had to track someone down?

— It's usually the other way around with me.

— Well I tell you this, if person A wants to find person B he finds him. Person B can't run away for ever.

— What are you talking about?

— Family stuff. You want to go back to my place? Let's fuck.

Charlie adored Rocky Sporgente's girlfriend. He loved the perfection of her limbs, the passivity with which she let him arrange her as he chose. He loved the smooth creaminess of her skin. He loved the sweet sad smell of her, like flowers that had lived too long. He loved the faraway look in her eyes. He had even learned to half-enjoy the violence that she provoked.

Charlie and RSG at Charlie's place. She thumbed through a

148

coffee table magazine. She wore a white silk blouse, a long grey skirt, a black shawl, black pumps. Her hair was held back with a clip (polyurethane, a gift from Charlie). The skirt had a slit in it, her legs. Charlie attended to the drinks. He played back the messages on the answer machine. She was bored on the couch. He walked towards her, she turned away from him. He stood behind her, she opened her legs, the skirt fell away. Charlie reached, Charlie touched, the rough touch that she had shown him to use. She tried to close her legs. He kept them open. And when he moved his hands away her legs stayed open. Charlie moved north. Charlie grabbed her tits. He squeezed the nipples between his fingers. He let go. He squeezed them again. She said, *That hurts*, in a voice that mixed curiosity and approval. He squeezed harder. Tender and bruised, they hardened between his thumb and forefinger. She was leaning back against the sofa as hard as she could. The back of her head butted into his belly. Charlie let go. She moved forward, away. He stopped her, still behind her, he pulled open her shirt, a button popped, bounced on the acrylic-topped table, rolled. *Don't bruise me*, she said. *Rocky will — Ouch!* Charlie was scratching her with his watch strap, long red marks into her perfect skin. Her throat was blushing. Charlie thought about vaulting over the back of the sofa, wisely decided against it, moved around. She was there, legs splayed, shirt and eyes open, nipples red, skin marked by his touch. *Creep*, she said. He hammered against her, knocked some of the breath out of her, ripped her skirt away completely. Marked skin, open shirt, white knickers. Pulled down the knickers to just below her knees, shaven quim. Charlie pushed her knees apart as far as the knicker elastic would allow, slid two fingers up into her cunt. Her head rocked, lips moistened, *Come on*, she said, *what more? This more*, said Charlie, trying not to feel foolish, his cock out of his trousers, forced it against her face, looking for her mouth, she moved away, he found it, almost, head against teeth, teeth open a little, nibbled on cock, Charlie forced it in, out again, can't rush this thing. Charlie pulled her towards him, moved her around, her elbows on the cushions, the side of her face against the sofa back, her arse in the air. What now? Charlie slapped the cheeks of her arse, red blushes to match the spontaneous ones on her throat; he wrapped his other arm around her waist, kept slapping, groans now, *yeses* and *no*s, he pushed her away again, bit her neck, her tits, her nipples, her navel, wrapped his mouth around her vulva, and now he repeated his watch strap move, he held her wrists with one hand, squeezing as hard as he could, dug the strap against her nipples with his other hand, pushed in again with his cock, and he was frightening himself with the force he was using.

— Stop! No. Stop!

He pulled away, let go of her wrists, relieved. This wasn't quite his thing. Sometimes he ran out of bad things to do — the watch strap was an inspiration that came from nowhere and troubled him — and call it squeamishness, but he didn't like to feel her hurting beneath him.

She looked at him, her grey eyes perfectly calm, knees jerking a little in and out, nipples scored by his teeth, cunt lifted towards him.

— Why have you stopped?

— You asked me to.

She shook her head, pity spoiled her beauty.

— You're crap. I didn't mean for you to stop. You'd have *known* if I'd wanted you to stop. I just like saying stop. I like the *word*. You're old. I'd have thought you'd understand these things.

Charlie ran his fingers along the red channels in her skin. Scornfully, she shook him away.

— What then, said logical Charlie, surprised to find his trousers had migrated all the way down to his ankles, would you say when you want me to stop? If it's not stop, I mean.

— You're so literal. I don't know, I'd say something like, I don't know, you think of something, a word, what's your favourite word?

— Favourite? Polyvinylchloride.

— Well that's really stupid. Look, I'm getting bored with this, just a one-syllable word, something ordinary, like green.

— All right, said Charlie, feeling contrite and amateurish and far from suave. Green.

They started again. He fucked her. Not forgetting, from time to time, to hurt her.

9

IT WAS SO easy.

— What are you up to Leslie?

Thursday afternoon. Leslie in his office, dusty surfaces over-crowded with papers, desk empty apart from scented floral lining paper and one paperclip wedged at the back of the top drawer, because Leslie had not yet dared to impose himself on his dead mother's world. When Charlie entered the office — quickly, not knocking, a useful habit — Leslie was cutting items from a magazine. He looked alarmed, a pornerast's fright, shuffled the magazine off to one side. He gripped his nostrils, blinked at Charlie.

— Up to? I . . . It's my journal.

He held up the magazine to show Charlie the cover, helpfully pointed a finger to the top so his cousin would find the innocent title, *Ancient History Digest*.

— Mind if I sit? (Charlie sat.) Mind if I make myself comfortable? (Charlie made himself comfortable, feet on desk, arms folded behind head, cigar wedged into the corner of his mouth.) You don't I suppose know any better than me where Howard is?

— No. A terrible thing. It's the children I feel sorry for. And Vivien of course too.

— Of course. Terrible. Absolutely.

— Would you like something? Coffee? Tea? Or (futile attempt to appear rakish) maybe something a little stronger . . .?

— I'm all right as I am. Never drink in work hours.

— Of course not. Didn't mean to suggest . . .

Perky Leslie, desperate to please, looked like an untidy schoolboy with no talent for acting who had been reluctantly drafted in to play an old man in the school play — wisps of

151

grey dandruffed hair, freckled pate, suspicious little grey eyes blinking hard behind industrial-strength glasses, long nose that cut his continental moustache in two, unlined face unblemished by the broken veins of sin.

— Busy, are you? Lovely day. While Howard's away I suppose we just have to go on with things, what do you think? Push on, maybe even, dare I say it? take some advantage . . .

Leslie nodded, briskly.

Why had he been skulking around on telephone calls to Bristol? He hadn't been contemplating dreams of being his own man this late in his life? Had he?

— . . . Here's a coincidence for you. I met a girl yesterday, Georgia. Saul's daughter. Says she chats to you on the phone sometimes. I wonder what you chat about.

Leslie blushed, as Charlie knew he would. He took off his glasses to clean them with a handkerchief the same grey colour as his skin.

— She's my niece.

— You talk about family, I thought that was it. Lovely picture. You bounce her on your knee, she tells you her stories, you tell her some of yours? Sit around the campfire swapping family recollections? And jokes too, I expect. How does that old one go — Why is Saul Cohen like a dog? Because he pisses on the family tree.

— I haven't told her who she is.

— Speak up. You're whispering.

— I haven't told her who she is.

— So what then, *Uncle* Leslie, have you told her?

Charlie was enjoying this. It was like shooting fish in a barrel, bullying Leslie into shame and worry, but still, all the same, he was enjoying it.

— Nothing.

— So what do you chat about? The weather?

Leslie looked around wildly, as if somewhere, as yet undiscovered, the shade of his mother hid, ready to take him in her arms to comfort him or, maybe, to strangle him.

— Chatterjee. What do you talk to *him* about?

Leslie showed some relief. To this question at least he could answer truthfully.

— I don't know any Chatterjee.

Charlie inspected his weak cousin carefully. Finally he nodded. When he spoke he was able to keep the contempt out of his voice.

— So you're doing this on your own. That's very brave.

Leslie would know what he meant by brave. He meant stupid.

— Your mother, said Charlie softly, would be very unhappy with you.

That hit. Leslie flinched, blinked. Looked pitifully small. He would whine now.

— What do you want with me, Charlie?

Good question. What did he want with any of this? Sometimes he was convinced that all he wanted was the big juicy coup, that special moment when the future was his, no one else's, and he would wave his magic hands, a puff of smoke, up in flames the company and his family and all the unnoticed things he has done for them, smouldering ashes; and when the smoke clears he is gone, standing puffing a cigar on the deck of a banana boat coming into the harbour at La Paz, a new snakeskin around the brim of his panama hat, the millions he has made converted into gems shining in the pockets of his linen suit, Rocky Sporgente's girlfriend beside him, or maybe not, cute little Conchita the harbour-master's daughter awaits.

— Call it co-operation.

Or else he remained the good son that Howard's betrayal has shown him to be. The one that stayed, righteous sweat on his righteous brow. Roll up his sleeves, oil the unwieldy corporate machine — pistons going, smear them with lubricant that drips like come — fix its rotten parts and turn it into something he could pass on to his unborn children, turn it into something his father and his uncle would be proud of.

— Leslie. Believe me. I want this company to be something good. Things have drifted. I want us to build something here, get rid of the dead weight, make something we can all be proud of.

Leslie sniffed. His face put on its hunted expression. He wiped his dry lips with an eczematic hand.

— I thought I could count on you. Are you with me?

— I am not my mother but I do what I can. But Charlie. Please. I hear something.

— What do you hear? People talking? A fire alarm?

— Something a little bird tells me. You know Harry Wiener, a good man, friend to our family, knew your father, knew Howard's father, used often to come for drinks courtesy of, bless her, my mother. He knows the business, loves the industry. Served as titular head of the Worshipful Society of Horners. Wonderful man, supporter of many charities. He's looking to expand, I hear it from a little bird, he might be agreeable to showing some interest—

— We're not selling off the company. Unthinkable.

— Of course not, of course not. We love our company. Awful thing to think about.

— You been talking to Wiener? Is that what you've been up to?

— Generally only. We meet from time to time, functions, we shoot the breeze, as they say. Not about this.

— You sure?

— What do you take me for? A viper?

— Do me a favour. If you do talk to Wiener, try to find out what he's thinking, that's all.

— Yes?

— Just get a sense of it. Pass it on if you hear anything or if a little bird tells you anything.

— Absolutely. Of course. You can count on me.

It was almost too easy. Just toss some fervour into your voice, look like a visionary, and Leslie would step humbly on to your team.

— Have you seen this?

Charlie brought out the photograph. Leslie reached for it and Charlie held it away just out of reach.

— Are you their enemy, Leslie? Are you trying to destroy everything they ever worked for? And your mother? Think of your mother, Leslie.

Leslie bowed his head, humble before the great men.

Charlie looked at his watch — and within the moment that it took for the platinum hand to tremble from one second to the next he had calculated that if he spun this out just a little

154

longer, and kept Leslie with him for the walk to the car park then there would be no fraction of time that he would have to endure on his own. The second passed, he chided himself for thinking so defensively, he shot his cuffs, he moved the cigar from one corner to the other with his tongue. Howard would be found, eventually, and Charlie's position would be strong by then. Leslie was in his pocket. He had once been in Howard's camp but that had changed since Howard ran away. (Who knew why? Who cared?) Dinah was more problematical. But Howard was the key and he would be made to fit — kicking and screaming with sweat dripping down his stupid bald head, or, better, with a smile on his lips, a song on his heart, tying the noose around his own neck as if it were the gayest of company ties — into Charlie's plans.

The only problem was, Charlie didn't know what his plans were. Charlie was congratulating himself on keeping things moving; and he was condemning himself for not making a decision as to where they should go. Growth or destruction. Either would be absolute and grand and irrevocable. The company torched and plundered, up in flames like the plastic combs the Edwardian cuties wore in their hair. Or rebuilt, improved with the labour of the good son's hands, perfect, ideal, moulded to his vision which would be identical to the vision of the company saints. Charlie let go of the photograph. He angled it on the desk so the faces would shame his sneaky cousin, and he crossed his arms behind his head.

10

CHARLIE'S FINGERPRINTS FADED: *grease whorls disappeared from his father's face and his uncle's groin, leaving
behind only a little more damage to the brittle chemical surface.
The photograph rested on the table and Leslie took the opportunity to hold it in his hands.*

— Excuse me, said Leslie. This photograph, Dinah was looking for it, at, the — my mother's — funeral. I wonder maybe I
should deliver it to her.

Charlie showed no sign of hearing him. He was involved in
private thoughts.

— This photograph? I can take it to Dinah?

— Tell me something, Charlie said. Tell me, what would you
like more than anything?

Charlie's face leering suddenly at him. It was a face that wore
its appetites on its sleeve, so to speak. And what a question.
There were many things he would like and he didn't know
which would come first ahead of the rest. He would like the
respect of his family. He would like the love of abandoned
Vivien. He would like to live in Israel before he died. He would
like to feel one day the absolute assurance that he was his own
man. He would like his mother still to be alive. He would have
her dead twenty years earlier. He would like to be rich. He
would like to be the magnificent benefactor of charities.

— Say it. The thing that would make your life special. Do
you want to be rich? Is there a lady you have your eye on?
Would you like to be able to draw?

Charlie sometimes was a mind-reader. Charlie had always
scared him. It was partly the brute stubby power of Charlie's
body, the lubricious things he said. But it was also the fear he'd

156

always had that Charlie owned the ability to look inside other people's heads, most specifically his own.

— Draw, no. I have never felt the urge to draw.

— Power and position and wealth, then. And pussy. How do they grab you?

Leslie bowed his head. Charlie knew the answers to those questions. He didn't need to say anything.

— I can give you those things, Leslie.

Leslie knew he could. He waited, head bowed, heart fluttering, for Charlie to go on. He didn't dare say anything.

— You can be second top dog at the company. Lots of money coming your way. You'll have to beat the ladies off with a stick. You'll be able to satisfy any whim you have. You'll be able to — come closer, these are magic words, aren't they? — *give money to charity.*

Yes. Surrender. Yes. Charlie knew. Charlie had the power. Listen to Charlie. He had been a fool to think he could scheme on his own. He had tried, in the first flush of his orphandom, to accumulate extra shares. Not to sell them to Wiener, not necessarily, but just to make his own position strong. And wherever he'd turned, whoever he'd called, Charlie had got there ahead of him.

— And for all this, what?

— For all this, my friend, very little. Back me up in my plans for the company. You know you can't do your piddling little schemes on your own. Trying to buy shares, make a deal with Wiener, it would never have worked, you know that. I'm sorry, I don't mean to put you down, but it had to be said. Back me up in whatever I do. If we sell, if that's what we decide to do — and don't worry about the land, it's worthless, polluted, you can see the surveyor's report, I'll take care of it — then we'll just split it straight down the middle. If we sell, I hope we don't have to, but if we do, we'll be able to squeeze half a million out of Wiener, minimum. Think of it, the machinery, the order book, our good name. Put your shares behind me. Help me persuade Dinah to sell her two per cent. Then maybe after that we won't have to sell, we'll make something grand. Fuck it, we'll make something special out of this. And one other thing maybe, if I need it, just one more thing. Some of the pension money, short-term only, it'll be our little secret, just so I can centralise the shares to do the good things I — *we* — need to do.

Their little secret. Leslie was not accustomed to secrets. Up till now he had successfully accumulated four secrets. A taste for Swedish magazines of busty outdoor girls that he used to hide from his mother. A liaison with a married Scottish woman who liked to call him Jacques because his moustache reminded her of France. His regular contributions to old people's charities and Israeli causes, always given anonymously because, as the Talmud says, the most virtuous form of charity is performed without show or hope of reward. Freda's last unsatisfied demand that she be buried in a coffin that she could escape from. Now he was about to have a fifth secret and he had never been so thrilled and so scared before.

— What do you say? I guarantee you it'll all work to your advantage. Between you and me (Charlie leaned forward, a tempter, a seducer, a guide to the promised land) I say, fuck Howard.

— Yes, said Leslie.

— What's that? Speak up. I can't hear you.

— YES! said Leslie.

They walked out of the building together, into the car park. Leslie waited along with Charlie for his driver to come. A nursing home ambulance passed by on Western Avenue and Leslie felt sad for all the old folk inside, without hope or warmth or companionship. Charlie asked a final question.

— Tell me. I don't know if you know the answer to this one but maybe you do. I don't know.

— Yes?

— It's a sort of ancient history question.

It was not like Charlie to hum and to hah. Normally Charlie was the perfect social engineer and Leslie just one of his many captive machines. Leslie waited for the question with some concern.

— Cast your mind back. You were there before us all. I joined the company. Then Howard joined the company. I was doing

158

all right here. I may have ruffled a few feathers but still I did my job well enough.

— So?

— I want to know this. Why Howard was made Managing Director ahead of me. That's what I want to know. Leslie. I have to say. It rankled. I had seniority. I didn't do anything wrong. It was Freda's doing. I know that much.

Leslie was relieved. That was an easy one. He had been expecting far worse.

— You married out.

— Excuse me?

— You married out.

— Out of what?

— Howard married a Jewish girl. You didn't.

A momentary glee to see Charlie so discomfited. Leslie wished he were clever enough to use this moment for his advantage.

— That's it?

— That's it. Listen. About this photograph. I may give it to Dinah? I'm seeing her tonight.

Leslie had always known that he was not an observant man. He recognised that things passed him by that other people would notice, that there was a world of useful information that somehow he always failed to pick up on. But now, despite the fact that he was not a good detective or observer of other people's private worlds, he saw something huge going on inside Charlie's mind. Like two dinosaurs facing off against each other, or Rome warring against Carthage, or the continental plates that so interestingly made up the earth shifting and smashing against each other, there was an elemental struggle going on inside Charlie, behind his lascivious face, a battle to the death that would shake all of their lives: until, an irrevocable decision had been made, only one dinosaur was standing, blood dripping from its sharp teeth, Carthage in ruins, the two continental plates had finished crashing against each other, the final volcanic eruption, and as Charlie's car driven by Charlie's sinister young driver sped dangerously into the car park, the seas were calm again, a new country had been made.

— Take it, said Charlie. Give it to Dinah.

II

LESLIE DROVE SLOWLY along Finchley Road to the restaurant where he would meet Dinah. The photograph was carefully in place inside his breast pocket. He and his mother often used to go to that restaurant. Sometimes just the pair of them, long peaceful dinners where they would chat playfully to one another almost as if they were romantically involved. (No one else had known the playful side of Freda's nature and that had always saddened him.) Sometimes Bertha joined them, always complaining. It was shocking that his sister was now in an old people's home. The years were used up so fast and clay returned to dust. Maybe one day, before his time came, he would learn how to escape from his mother's pull and maybe one day he would find a way to push aside the weight of the memory of his mother's funeral. He still couldn't believe that the day had been real, her ebony coffin lowered into the ground, the first sods of earth thrown on top by his trembling shovel, that his mother now rested underground inside a brown bag in an ebony coffin.

Leslie tried, he should have tried harder, he hated himself for not trying harder, for being too tired and fearful and sad to break anyone else's rules, to implement his mother's penultimate wish. Efficient Freda, the dry little widow, had a secret terror of being buried alive. She confessed it at the end, on her death bed. She had made him promise: No bag, a coffin with a hand-bell inside it, airholes too; and he had promised, head bowed, eyes finally wet with tears: *Of course, Mother, I shall.* And then she had made him promise to get married, it no longer mattered to whom, a Jewish woman it no longer had to be, just do it, become, at this late stage in your life, a man. *Of course, Mother,*

160

I — And then she had died, a last smile, a tremble of her closed eyelids, nothing more. And on the next day, they had laid her to rest, everyone there but Howard, Howard the Unforgiveable, who had not had the decency to attend his aunt's funeral, it was a crime that could not go unpunished. His mother inside a wooden coffin without any airholes, a handbell pointlessly smuggled in to rust next to the bag they'd wrapped her body inside.

Enough. He could torture himself for ever with the image of his mother trying to escape from her stifling coffin. He had arranged to see Dinah tonight, and even if she showed no desire to see him, still his company was good for her, it was important for old people to have excursions.

They met at the kosher vegetarian restaurant where the prices were good and even though the service was slow it was not as surly as some. Leslie tried to interest Dinah in different subjects: in her bridge circle, in the surprising information that the man who presented that crime programme on the telly was Jewish, in the possibility, maybe, just for her to think about, that maybe she should contemplate selling to nice reformed Charlie her shares in the company, and, in a careful roundabout way, in Vivien, but she showed no sign of interest in anything; and so he had been forced to bring out the photograph, much earlier than he had intended, before their first course had arrived.

Dinah looked, as if for politeness, at the photograph in front of her. What was going on inside her head? She wasn't touching the picture, maybe not daring to. She examined around the edges first, the workshop clock on the wall fixed for ever at five to five, the metal trays containing engineers' things, the table at the side.

— It is the right one? The snapshot you were looking for?

No answer. Peculiar expression on her face. Hunger maybe, he should have taken her somewhere where the service was faster. He had expected more emotion from her, pleasure, gratitude even, she had asked them all to search for it. Twice she had asked him. Once at the funeral and once several weeks before that. His mother had sat up with him one night before her final illness took her, and they had gone through all their albums together without success, but it had been an enjoyable evening none the less.

— Auntie? Excuse me. They are trying to serve.

Dinah apologised to the waitress with a graceful smile, moved her hands away from her place setting. The minestrone looked thinner than usual. She was about to speak. She would thank him now.

— Where did you find it?

— Charlie. He had it. I explained to him how important it was to you.

— You mean Lawrence.

Only Dinah now was left who still called Charlie by the name his parents had given him. Lawrence had been rejected long ago as too Jewish, too suburban. That at any rate had been Freda's judgement. Howard had had an original theory about why the name-change but Leslie no longer remembered what it was and he was glad of that.

— Charlie. Lawrence. What's the difference? He had it. I took it for you.

— Thank you. Yes. I'm very grateful.

Leslie's pride was being ruined by disappointment. She should be showing more gratitude. They had all been asked to find it and only he had.

— Leslie. Do you believe happiness is important?

The question worried him. When his mother had turned philosophical it had been a sign of her impending end.

— Do you feel unwell, auntie? We should change tables maybe. They wouldn't mind. There is in fact a through-draught.

— Look at his face. At Danny's face. And Sam's too, he's happy. I think that must be why I wanted this picture so badly. I've been trying to picture that expression and I haven't been able to. Tell me. Have you ever been happy, Leslie?

— There's a question. A big question. One of the Greeks had some interesting things to say on the subject. Not Plato. Diogenes maybe. He, when — Aristotle? No. I don't recall.

She leaned forward, she still looked fit, much younger than her age. Spry, perhaps, was the word.

— Do you think there is even any such thing?

— I have to say, auntie, that what you are asking is philosophy. It is not my specialist subject. I remember that Charlie once said the trick about being happy is to pretend that you are.

— Your cousin is very foolish. Please. I would like to go home now. I really am not at all hungry.

She pushed aside her untouched soup plate. Leslie suggested trying another restaurant but she seemed to want neither food nor company, at least not his. He arranged with the waiter for a sandwich to be made up for her in the kitchen and Dinah submitted to take it, wrapped thoughtfully up in grease-proof paper. She told him she would take a taxi cab home and Leslie said he wouldn't hear of it.

The traffic south was surprisingly busy. Neither seemed able to think of anything to say to the other. He walked her to her door and they kissed good night and Leslie drove home as slowly as he could. He had driven this route so many times with his mother. Freda criticising his driving, telling him to go faster, to go slower, to drive less jerkily, to watch out for madmen up ahead. There was a young woman who reminded him of his mother standing alone at a bus stop. He didn't know why she reminded him of his mother, there was no similarity of age or looks or even skin colour. Leslie stopped the car, lowered his window. It was dangerous for a woman at night in the streets. He offered the young woman a lift. She didn't look at him as she answered his consideration with abuse, terrible words, some of which were new to him.

He arrived outside the block of flats, his usual parking place. He couldn't bear to go in yet. She would haunt him for ever, and the ghost of her face would always wear that last terrified expression when she made him promise to bury her in a coffin she could escape from. The doctor had said it would be good for him to see her body, at rest, at peace, so he could remember her that way, not when she was losing the struggle to go on living. It hadn't done any good, just made him feel worse. He knew that he had to reorganise everything in the flat. He had tried to sleep in her bedroom but that felt like incest. He still made two cups of tea in the morning, two mugs of cocoa last thing at night. Sometimes, even, he stayed true to the habit of the last bad months and brewed up one of his medicinal

Specials. It was only when it was ready on the tray, honeyed and lemoned and hot, that he would realise that there was no patient to deliver it to. Down the sink it went. Then wipe down the tray and return it to the cupboard of unused things. He should get married, perhaps that would allow her to rest. A man was not complete on his own.

He wished he could find more reasons to visit Vivien. She did everything so beautifully, with class. She had troubles of her own, but maybe that was a good thing: she needed someone to lean on, he needed someone to lean on, and there was that way that she looked at him, as if she knew what was in his heart and didn't condemn him for it.

It was late. The only light on in the mansion block was his hall light, kept on to discourage burglars, and also as a first rebellion against his mother who had hated unnecessary use of electricity more than anything except, maybe, homosexuals. The photograph of Danny and Sam was on the passenger seat, overlapping the permanent impression of his mother's narrow bottom. Dinah had forgotten to take it. He couldn't go in yet, he felt as if he would never be big enough to enter his flat again.

Leslie started up the car, it jumped forward, stalled. This problem had been with him for a while. He should get it looked at but all mechanics are thieves. The car started second time. He moved away from the curb without indicating or checking in his mirror, he didn't need to, his mother was looking after him, he couldn't come to any grief. He drove the familiar journey back to Dinah's house. What was Charlie planning? He was up to no good, that was certain. While Howard was away the mice would play. He had been foolish to think he could get things done by himself with Wiener. He wasn't strong enough for that. Whatever Charlie was planning he was part of it and Howard was not and that was enough for him. The black girl was no longer at the bus stop. Leslie sadly contemplated — no blame attached to him, who had at least tried — the urban disasters that had befallen her.

There were no lights on in Dinah's house. He must have been sitting for such a long time outside his flat. Leslie left his car, it was a nice quiet neighbourhood this, maybe he should move here, perhaps even move in with Dinah, this house was too big for an old lady by herself. Or Vivien maybe. Alone in that nice house with two difficult children to look after. He could make himself useful there. People should be useful to one another. If Howard never returned, who could say what might develop? Leslie lifted the horse-head door knocker and quietly lowered it again. It would be unfair to wake up the old lady. He knew from bitter memory how much trouble old people had with their sleep. He pushed the photograph through the letter flap and walked slowly back to his car.

FOUR

GRANDMA

I

SHE HAD BEEN dreaming. A Buckingham Palace garden party. Grey glamorous moments on a charmed summer day. Country vicars lifted their heads from their game of croquet to admire a caravan of underprivileged children in wheelchairs rolling across the lawn. Large ladies in pale suits nibbled little tea cakes covered in icing. Royal princes dressed as Elizabethan privateers greeted each other with a flourish of feathered hats. She was sitting under the shade of a cherry tree. She was balancing a cup of tea on one knee and a china plate of barbecued chicken and potato salad and coleslaw on the other. The Queen was beside her and was presenting her with a telegram, in person. The Queen was asking her a question but she couldn't hear what it was because somewhere a kettle was whispering and she had forgotten the term of address she was expected to use (Your Majesty? Your Highness? Ma'am?) and she was feeling guilty that she wasn't really one hundred years old and there was an awful whistling in her ears.

A curtain pulled apart. The Queen became a dark cartoon bunny on folding velvet. Greyness. The slow settling shapes of her familiar room. A cold top sheet that tasted wet. The whistling of a kettle that came from downstairs. Dinah eased herself out of the bed. She sat for one shivery moment on the edge and delicately shook her head to disperse some of the customary cloud. Then, carefully, she stood. She wrapped herself in her kimono dressing gown and stepped inside fluffy blue slippers made odious with the faces of kittens (given to her by someone she'd forgotten who was trying to get into her good books). She didn't bother with her glasses or the light switch. Darkness has never troubled her. The whistling continued. A kettle.

169

She took the stairs slowly. She kept her left hand on the banister rail and held her dressing gown closed with her right. She dealt with each step respectfully, on its merits. Right foot down first, then left foot to join it. The whistling wasn't getting any louder, which was strange. She counted down each step. Thirteen. Twelve. Eleven. It was cold in the house. Ten. Nine. She worried that she had left a downstairs window open. That was an invitation to get burgled. Eight. She was glad that the kettle had woken her to this. She would double-check her ground floor security. Seven. She felt the harsh need to cough. Six. Another noise, sinister, a metallic flap, this could truly be a burglar, she must be silent.

She lifted her hand off the banister quickly to cover her mouth. She coughed a small bolus of mucus into the side of her fist. And lost her footing. She missed the fifth stair from the bottom with her left foot. A moment of sudden panic, blood thudding in her ears, a lurch of her heart, as her left foot foundered on air and her right ankle began slowly to twist. She reached out with her wet left hand for the banister rail but she missed it. Her right hand waved up into the air to try to grasp back her balance. Her dressing gown fell open and she remarked out loud to herself how cold it was. She repeated the sentence three times and it surprised her how long it took to tumble down the final five steps and spin on the bottom one and hit the hall floor with the side of her left knee and feel her weight crumple down over the top of it. Her chin hit the carpet hard and she tasted blood in her mouth. The whistling, though, had stopped, and she was glad about that.

Things became pleasant and strange. It was very comfortable to be lying in that position, her left leg broken at the knee beneath her, her mouth and nose pressed intimately against the nap of the carpet. She had inadvertently discovered a new way of lying. She instructed herself to remember this position for always. The only disappointment was the narrowness of her vision. All she could see were the two shades of carpet green weaving their paisley way across to the front door; the bottom third of the door, with just the lowest edge of the wire grille of the letter cage, and something inside it, white border, grey shapes fuzzy and blurred; the narrow curved legs of the hallway table (telephone remotely at the top, the hidden silent summit of a faraway mountain) which tapered into feet like monkey paws; and two Japanese boys in loose black trousers and white singlets. They had bare feet and harsh savage expressions and their black fringes of hair were stuck with sweat to their smooth foreheads and they were dancing for her. It was a sophisticated dance. She would have liked to join them in it but she was too comfortable where she was and anyway they didn't seem to require anything of her quite yet but to watch. So she watched.

They made intricate movements and some quite wild ones, and the dance sometimes took on the urgency of a fight and sometimes the sad innocence of a lament. They twisted together, at each other, chests grazing, knees jerking out but never colliding, and they gently corkscrewed to the floor and when it seemed as if they were about to kneel they sprung up into the air, hands across chests, elbows out, legs kicking wide apart, and for a while they floated. They landed on their toes and slowly flattened their feet and lowered their chins and they bowed to her from the hips, and she would have liked to applaud but she couldn't do that.

They sat cross-legged on the floor either side of her head and they whispered endearments in their own language, which she could understand, and they stroked her hair. They marvelled at its fineness and its whiteness. They laid strong tender fingers on her cheekbones and praised the classical beauty of her features. One (she thought this one was called Koko) was the gentler of the two. He wiped away the blood from her mouth with his fingertips and scraped it away from where it had congealed at the corners with his sharp nails but he made sure this did not hurt her. He sang a sad lullaby to her from his own country. The other (Yoshomoto) stroked her feet and legs. She wanted to tell him to stop. She didn't want him to be appalled by the horror of her varicose veins. She didn't want to feel his sharp hard fingers poking into the tender places of her bottom. But Koko consoled her. He knew what she was feeling. He raised a finger to her, nodded as he lifted it to his closed lips. She blinked back at him and he understands. Koko stroked her hair and her temples and her neck and sung his lullaby to her and reached over to massage her shoulders and he smiled to her and made her feel safe, and Yoshomoto rubbed at her thighs and stroked her broken knee and followed the curled ropes of her varicose veins and spread open the cheeks of her bottom to trace the dangerous outline of her haemorrhoids. And Koko lifted her chin off the floor and cradled it in his hands and leaned down with perfect dancer's control to kiss her on her forehead and then her cheeks and her nose, and while Yoshomoto explored parts of her body in which she thought sensation had long ago died, Koko finally kissed her mouth, and now she could see most of the door, there was the greyed, wrinkled back of an old photograph in the letter cage, and she couldn't remember the last time she was kissed on the mouth except by accident.

Then the two Japanese boys curled like fauns beside her and she slept for a while. They awoke her because a procession was going past and they didn't want her to miss it. Koko and Yoshomoto crossed their arms across their chests and nodded at the grandeur and severity of it. It was her funeral, she realised that quickly, and it was grand and severe and even splendid. Horseshoes of white lilies on the backs of big German hearses. Sorrowful men in frock coats sitting upright in the front. Sobbing women behind them dabbing wet eyes with white handkerchiefs. Many cars. Many grievers. She wanted to tell them not to be so sad. She wanted to shout out about the happy place she had found with her two new friends, but Koko warned her with a look to be silent so she was silent. She recognised them all in the hearses. The children dressed up in thick mournful clothes but smiling and happy despite it all. She saw Lawrence who calls himself Charlie and most of his wives, and Howard was talking to him and they each had a black-jacketed arm over the other's black-jacketed shoulder, united at last in shared sorrow. She saw little Simon, covering his senses with cotton and wool and plastic. She saw Ruthie and Vivien and Leslie and Bertha. She saw her sister Esther with strong fighting Sam and their first son, Maurice, returned from the East. And she saw Freda and Jack and Sarah and her parents and even Daniel, her long dead husband, with no scar of fire upon him.

And behind the long line of hearses, there was dancing and singing and she was glad to see Koko and Yoshomoto leading

172

the songs, and there was a monkey with them, dressed up in military uniform with red piping and gold braid, a girl monkey, with a neat parting of hair down the centre of her pretty skull, beating on a silver drum. She saw it all and she would have liked to clap her hands together at the joy of it. She approved of the sights, the flowers and the blackness and the clothes, and the solemnity of it, and she liked the dancing and the monkey and the singing, and she was glad that Koko and Yoshomoto were in an important visible place and everyone would know how happy the two Japanese boys had made her.

The monkey banged louder, louder on the drum, and she admired the rhythm of it, it was like the rhythm of her own blood. It made her happy that she had finally realised the central blood truth of drumming, but people were shouting at it, the monkey's pretty face was getting confused, and Dinah started to feel terribly sorry for it and quite defiant for it for its own sake. And that was because the monkey was a pretty, gaily dressed thing that was hairily graceful in all of its movements, and it had only wanted to celebrate and make noise in her honour. They were all shouting at it, men and women and children, screeching the monkey's name, which was her name too, in tyrannical family voices. Some got out of their cars to scream but others screamed behind their closed tinted side windows. And the monkey beat on its drum, no longer such a satisfactory rhythm — ugly — irregular — and the people beat on the doors and the windows and the walls. They shouted out grandma and auntie and Dinah and they were confusing the monkey quite entirely and Dinah didn't like this, it was disrespectful for them to be speaking to the monkey as if it was her.

Then the door was beaten down. The wood splintered near the bottom and then there was another kick and the sole of a man's shoe was suddenly caught there, stuck, not far from her face, and she laughed to see it, it looked so ridiculous and male, wriggling like a dying fish. It had been caught at the ankle, and there was a spot, just at the bone, where the beige sock had torn and the thin skin of the ankle had been ripped lightly open, and each silly wriggle was rubbing the cut further open against a splintered door plank. Koko and Yoshomoto got up to disentangle the man but she told them no. It served him right, disrupting her funeral and being so cruel to a monkey.

It was a mistake to talk. It hurt her chest and encouraged the enemy. A second hole appeared in the middle of the door, and then another just beside it. And two unfamiliar faces briefly appeared at the middle holes, just at the same time as the shoe went from view (accompanied by a scratching sound and a loud curse, followed by a long petulant moan). The faces disappeared and an arm took their place, articulate fingers dancing out a message she didn't understand, finally alighting on the key beneath the door knob and turning it, and then sliding the security chain out of its housing and twisting the door knob. The door pushed open, and she watched its swift path towards her and she enjoyed the sensation of it lodging against her forehead with a reassuring thud.

It got a little confusing after that. The hallway became so filled with people that there wasn't any room left for Koko and Yoshomoto or even Dinah the monkey, no matter how agile she was. People fussed around and called other people on the telephone and whispered conversations and moved things away from where they were supposed to be, and they all made very curious shapes if you looked at them from above, which Dinah chose to do for a while until being so far away from herself made her scared. There were policemen in blue uniforms and blue helmets and firemen in red uniforms and yellow helmets and representatives of her family no longer in funeral clothes, but apart from Leslie weeping in a corner and Simon standing beside her there was no one who had any business being here. Howard wasn't here, and neither were Esther or her parents or fighting Sam or Freda, nor even, and this surprised her more than anything, dead Daniel.

She wanted everyone to go away and she told them so. But even as she was saying it she was beginning to realise that without Koko and Yoshomoto to look after her (why had everyone been so horrible and scared them away? the thought of it made her want to cry, so she did) she would be in some trouble trying to get back to her bedroom, and the telephone was ringing and the kettle was whistling again, and she started to have serious reservations about the comfortableness of her lying position because her broken leg that had been folded so neatly under her was burning with pain, and her spine and her bottom were throbbing, and her face was aching as if it had lost a layer of skin, and she needed to go to the toilet very badly, and she missed her Japanese dancer friends very much, and no one would admit to having seen them. (Instead they talked in hateful whispers — why did they think it necessary to talk in a language she didn't understand? — words like inshock and tinnitis and brokenleg and poorloveshesdeliriouswithpain and isitthestartof-dementia dyouthink?) And finally they bullied her body into a straight unbroken line and they strapped her body onto a tray and someone stroked her hair with an unpleasant touch and someone covered her with blankets that hurt her with their pressure, and someone else slid a photograph inside her kimono next to her skin, and looking at the broken door that she would never see repaired made her cry some more. But Simon made her feel a little better. He held on to her hand as the men in blue and red uniforms carried her out on a tray past lines of tea-clutching mouth-agape neighbours and he spoke to her as the shock of the breeze closed her eyes and made her feel shivery and young and very breakable. He told her that they hadn't gone away and she knew what he meant and so did he.

2

THINGS GOT TOO quick for Dinah. She was wheeled and rolled and walked and washed. She was dressed and undressed. She was filled and then she was emptied. Electricity and syringes were poked beneath her skin. Clear fluid went through a tube into her arm and came out as yellow fluid through a different tube. A handsome distracted man cut her leg apart with a knife and an ugly distracted man sewed it back together. A West Indian lady told her the secrets of God in song but Dinah couldn't catch what they were because of the unfamiliar accent. Her leg was encased in beautiful white plaster and then fixed into a dark metal apparatus with shiny edges that looked like a giant insect. They gave her forms to add her tentative name to and rusty wheelchairs to sit upon. They put her in a scratchy narrow bed and surrounded her with soiled blue walls and travel posters and outside they gave her cracked white corridors to roll along. And she would huddle in a corner of her chair, her plastered leg straight ahead, keeping her body as much to herself as she could, and she would watch out for Koko and Yoshomoto and even though she never saw them she knew they were there somewhere. And she would touch herself at night in the nostalgic cruel way Yoshomoto had shown her but it didn't quite work as it had when he had done it, and she had to learn to do it secretly, because whenever they discovered her doing it they would stop her. There were some who slapped her hands away and tied them across her chest. There were others who would gently hold them still while they tried to grab her attention by singing dull songs from somebody else's childhood. There was one who would add his hands to what Dinah was doing with hers but she didn't like that very much. Sometimes he hurt her, but on reflection she preferred his attentions to anyone else's, especially the simpering cow-faced woman whose yellow-grey hair was bundled up in the shape of fruit and whose unintelligent questions got in the way of her thoughts. They cheerfully told her she was a difficult patient. They talked in loud incurious voices around her. They discussed her decline and her withdrawal. They put her in a smaller room which smelled of dried blood and air freshener.

Simon couldn't come to see her. She understood why. Howard was never there and that was inexplicable. But Lawrence who called himself Charlie came. He came every day.

— Your name is Lawrence, she said to him one heavy morning after he had bored her into communication. Your grandfather. Your mother's father. You are named for him. Not some spiv bookmaker Charlie.

— Don't worry, she heard Lawrence say. Don't get excited. You mustn't worry yourself.

He had changed his name as a thirteen-year-old, announced to Sam and Esther the day after his bar mitzvah that he was Charlie now. It had hurt them at the time, hurt Dinah still. He had been given his name to honour her father, Esther's father, and his rejection of it was a sign of the way their whole history had been lost, buried in the foolish proud madness of Danny's family.

— Are you comfortable? Do you have everything you need?

She coughed for quite a long while, and under the cover of her coughing she swore bad words at her nephew and rubbed herself briefly with her hand which was not entirely with its proper feeling.

— It's a nice place here. Are you happy with everything? The nurses seem very thorough.

Dinah wanted Lawrence to go away. His noise irritated her. It wasn't hygienic for him to dig his nails clean with a pocket knife. He had never been a sensitive child. He had too much hair and he wore it lavishly long and Dinah despised him for it.

— Everyone says you're on the mend. Give me your hand. There. It's a lovely day. If you're good the nurses say you'll be

177

able to sit in the sun in a week or two. How's your leg? Does it hurt you?

Her leg didn't feel any part of her. Only her eyes and sometimes her hands and her imagination and her capacities for boredom and contempt truly belonged to her still. She knew what Lawrence was after and he wasn't going to get it.

— Has Howard been here to see you? Have you thought some more about your shares?

Dinah pulled her head away from her nephew's rough hands. The signet ring on his little finger had been digging into her temple. She wanted to tell him how great a man his uncle had been, despite his strangeness and his weakness and the tragedy of his end, but she didn't have much use for speaking and anyway she knew that Lawrence would only pretend to pay attention to her words. She looked at him and didn't like what she saw. Red face puffed up with over-eating and vanity. Short heavy body. The snappy spiv cut of his tweeds. That ostentatiously ginger tumble of hair styled like a sharper's. She shut her eyes and listened for the sound of a monkey chattering but she couldn't hear it yet. She heard it sometimes at night, when the woman next door had stopped praying, but the monkey was a mischievous creature and had stayed out of her vision so far, taunting her to find it.

— We could move you somewhere else if that is what you would like. Is that what you would like? It would cost more but I would do it if that was what you would like.

The monkey was teaching her something, she knew that. It had something to do with the value of mischief, and play. And agility too, perhaps. She envied the remembered movements of the monkey before she had been pulled out of her house.

— I'll be back tomorrow, darling. The same time. I'll see you then. Be good.

Charlie folded his pocket knife away. He put his coat with the velvet collar back on, leaned over the bed to give his aunt a kiss on her forehead and a signet ring rap on the cast of her broken leg.

— Look after Simon, she commanded. Simon is more special than the rest of you.

Before Charlie could respond to his advantage to Dinah's

sudden articulacy, she closed her eyelids again and rolled her eyes up as high as they would go. People didn't trouble her now she'd mastered the trick of keeping her eyelids still when she pretended to be asleep.

3

ON HER BEDSIDE table: a plastic water jug embossed with yellow stars; a Tupperware drinking cup; a red and black box of TenderSoft™ tissues with the next one ready and poised; a bunch of grapes in a rainbow-coloured fruit bowl; a small pile of paperback romances from the hospital library, their covers wrapped in cellophane to protect them from diseases and finger-dirt; an almost new lipstick from Freda's dressing table; a selection of silver jewellery; her spectacle case with its frayed blue cloth cover; the box in which for the last half-century she had stored what jewels she possessed; and a photograph in a makeshift white cardboard frame. Her husband and Esther's husband together in youthful glory.

When the dead outnumber the living then you know where your responsibilities lie. Dinah could no longer remember the birth of the impulse that had made her want this photograph, but it had obsessed her for a time, she remembered telephoning everyone who was storing anything of hers, she had gathered useless dead things around herself because she had hoped to find among them a photograph of her husband and his brother in their first triumph. She had thought at the time it was because Danny looked so happy, and that was a look she was no longer strong enough to remember on her own. But more recently, since her accident, she had wondered if it was because she had reached so near to death herself that the time had finally come to investigate the mystery of her husband's heart.

There had been a strange intense boy, a blond gentile with a sharp face and dark intense blue eyes, who had been her suitor before she was forced into an intimacy with Daniel. This boy was an art student at the Slade. He was filled with passion

180

and articulate with foreign names. He was a neo-Futurist, he proclaimed. He believed loudly in speed and power and war machines and free love, the limited beauty of the human body, the absence of the human soul. He liked Dinah because of her genealogy, he told her, a shepherdess, a goat girl, formed in fire in an eastern crucible of sun and desert. He would take her swimming at night in the pond by South End Green, he naked, she modestly wrapped up; and afterwards they would lie on the grassy ground, and he would throw himself upon her, bellowing out names of artistic doctrines. And she would resist, pushing him away, a desperate battle, until he would cry out and fling himself from her and shudder his body against the ground and pulse his semen into the unprotected Hampstead earth.

Daniel took no such liberties when she brought him there. Dressed in an ancient swimming costume that hooked over his skinny shoulders, billowed over his chest and trousered down to below his thin little knees, he would follow her across the cold grass, his naked feet nervous for insects and sharp stones and dead animal things. Until they had reached the pond, where Dinah would pull him complainingly onto the wooden jetty. She would smile, squeeze his hand one time, then dive off into the black water. She would look back at him from the consolation of water and see him there, on the end of the jetty, Daniel sitting down, his feet pulled clear of the sinister pond, his arms around his knees, his eyes shut, listening out maybe to the birds in the trees, the splashes made by Dinah's strong legs, the midnight cars off on the road, perhaps he was finding reassurance in the machine sounds of the carnival gypsies oiling their gear for the Easter fair. Then, before Dinah grew too impatient with him, too scornful, in this one place where she could assert herself, her element of water, he would slowly lower his narrow body into the enemy pond, execute a few noisy inefficient strokes and then, honour restored, pull himself painfully back onto the jetty. He would lie there, shoulder blades hurting against the wood, the breeze assaulting his shivering body, waiting for his future widow to be finished with her watery needs.

Then the painful rub-down with the towels, the brief difficult walk to their piles of clothes, the liberation of shoes, and into the car, which Danny drove back to Stamford Hill in a screeching lurch of gears. He could invent machines that no one had ever imagined before, he could repair machines beyond their makers' dreams, but he never could use them very well. Had he been happy then? Had she?

4

VIVIEN WAS TIDYING the grapes. She picked off the empty branches and dropped them on top of the mound of tissues in the grey metal rubbish bin. She centred the bunch in the bowl and arranged the loose grapes, stems up, in a circle around it. Then she returned the bowl to its place on Dinah's bedside table, taking care not to obscure the patient's view of her photograph. Vivien leaned down for her magazine. She found the competition page, glanced briefly out of the window where hospital porters in blue overalls were sharing a can of beer, lifted the heels of her feet away from her shoes, and attended to her anagram grid.

Dinah had learned how to watch without seeming to. She had overdone it at the beginning, rolling her eyes and trying on random smiles and frowns, with the result that people stole her blood from her and rolled her to the X-ray corridor where they cut away a patch of her hair, which made her cry, and fixed metal plates to her head and buzzed ugly colours inside it. She had learned now to be quiet. To keep her world separate. To admire her monkey with the same name as her on the few times it volunteered to appear, and she was beginning to learn how to coax it out of its hiding place. Dinah would hold her breath and close her eyes and slow her heart and she would make her self very small, and then when she was small and still and lifeless enough she would open her eyes again and sometimes there the monkey would be, spinning by its tail on the curtain rail, chattering to her of the things it had seen.

Vivien's visits were full of things unsaid. Dinah had learned a long time ago that Howard's wife never talked of anything that moved her. It was as if allowing the emotions to have words attached to them made them too unbearably strong. So she tidied things up and filled in puzzles and Dinah kept to her place and built a silent wall against the ugly world outside.

Vivien put her magazine down on her lap. She cleared her throat.

— They're moving you soon. They're very pleased with you. They're putting you in with three other ladies. That'll be nice company for you.

Dinah tried to see Simon in Vivien, or even something of Ruthie. The children were more from Daniel's side, and hers, and that pleased her. Apart from her wildness Ruthie was like Esther. She had the same womanly shape and carried herself in the same sunny-side-of-the-street way as Esther, who had been Dinah's sister, and was scarcely old enough to be a mother and a widow and dead. Lawrence was like his father Sam. He had the same body as him, but then it was the same body as Sam's father too. Strong man Jack with damaged lungs and a taste for gambling and dirty jokes. Simon was Daniel. That was clear enough to see. She didn't know where Howard came from. He was part Daniel and part something unknowably else. Howard was taller than the rest of them and skinny like his father, none of that family power the other men had in their muscles, and he looked old, he always had. He was polite as a boy, and he groomed himself well, but he'd always looked as if he was wearing someone else's clothes. He was removed from things like his father had been, and he made mistakes like his father. He had always refused to learn that Danny had died in Pimlico. Howard had heard it as Paddington when he was very young and something in him had decided never to change his mind over that. Long nose and grey face and a lot of black hair, like Daniel, except he'd lost most of it, something which Daniel never did or never had the chance to do; and Howard had frightened brown eyes, which for a long while, long ago, she had tried to soften and lighten. Why hadn't he come to see her?

— Do you have everything you need? How often does Charlie come to see you?

They'd said her condition had worsened. Dinah had heard them talking, mildly insulted that they didn't take the trouble to hide their voices from her. They'd talked about dangerous

184

things going on in her head. They'd argued about her blood. They'd said they would need the room for a new patient. They'd told Vivien that Charlie had given permission. They'd asked where Howard was and Dinah had heard Vivien tell them she didn't know.

— You look a lot better.

Simon was Daniel and Lawrence was Jack. More than Jack. He had a gusto that Jack never had. That at least came from their side. Jack had frightened Dinah in those early days with Danny. A small heavy man, weak despite his muscles, listening to the radio in the front room and smoking cigarettes and drinking beer. When he wasn't listening to the radio he was reading a socialist newspaper, lifting his face away from it only to order his little wife around and cough and make remarks. After Sam had married her sister and when Danny and Dinah were beginning their courtship, Esther had warned her about the father. *He is sick*, she said, *and as a result, cruel. You will have to make allowances*. Dinah tried to make allowances. She endured long wasted hours sitting with tea with the father, watching him cough, watching him spill his beer on his corduroy trousers and shout his wife's name. Watching him listen to race meetings on the wireless that Danny had improved with his occult gifts, and watching his round unhealthy face as it enjoyed the effect of those words he used whose meanings she tried so hard not to guess at.

— Come here, said Dinah.

Vivien had not anticipated anything so direct. She leaned forward, rested a light hand on Dinah's narrow wrist.

— Closer.

Vivien edged forward on her chair. Her face was a nose away from Dinah's. The magazine fell to the floor and that pleased Dinah.

— Fuck, said Dinah. Fuck and cunt and ca-ca.

Vivien nodded gravely, as if this was what she had been expecting.

— I think, she said, I should go.

— Ca-ca. Corduroy ca-ca.

— I'll come back tomorrow.

— Fuck. Fuckity fuck fuck.

Vivien moved the hair away from her brow. She tried to think of a suitable exit line and the effort brought out the lines on her face.

— I'll bring you some plums. They're in season.

Dinah had not enjoyed herself so much since her accident. She felt a pang of sympathy for Vivien but it was her own fault for not talking about things that were real. Why didn't she know where Howard was? Why didn't she talk about it? Dinah pressed mercilessly on.

— Shit. Schmuck. Timtum. Feigele. Haimishe. Zeiger zint.

She wasn't sure about the Yiddish. The language had always startled her. Made Danny's home even more forbidding, this ugly eastern vocabulary that seemed designed to describe penises and cucumbers in all their different stages. She had never understood the language's obsession with cucumbers.

— I don't like cucumbers.

— I said plums. Plums! (I'm sorry to shout.) I didn't say anything about cucumbers.

— Penises and cucumbers. That's what it's all about. Why?

— You should, said Vivien, and I'm sorry to have to speak like this, watch your language. You'll be with three other ladies soon. I'm sorry. Honestly I am. If you could see yourself you'd be — oh. I'm sorry. I just don't want you to upset anybody.

— And shit, said Dinah. Penises and cucumbers and shit.

TO LESLIE SHE didn't show herself at all. She didn't dare give him the encouragement. Sitting at her bedside, rocking himself in the metal hospital chair, pouring out his grief and his guilt as if she was an enormous vessel capable of containing it all.

— I have brought you this.

This was how he always began. A foolish smile, a wet kiss, then the step back, and his hand stretched out to wave the thing in front of her face.

— It is not much, a little thing only, but I thought you might like it.

He would put it on the table. His guilt offering. A bunch of flowers, an array of grapes, a box of chocolates, or something plundered from Freda's dressing table which he assumed Dinah wouldn't recognise but she always did: an almost new French lipstick, an eau de cologne spray; this time it was a powder puff in an antique-style silver compact case.

— I blame myself, he said.

This is how he always went on. He sat and he rocked and he blew his nose as if he was intent on tearing it apart and he blamed himself for everything. He blamed himself for her accident (if he hadn't slipped the photo through her letter box . . . if he had only reminded her to take it in the car . . . if they . . . if he . . . if only . . .); he blamed himself for something to do with Freda's death or it might have been her burial — even when she tried Dinah was never able to disentangle the meaning that lay soaked somewhere in his damp awkward words; he blamed himself for not having been a perfect employee of the company; he blamed himself for not having been a better son to poor martyred Freda and a good brother to poor unwanted Bertha;

he blamed himself for something to do with Lawrence and the business and a man with a name like a sausage; he blamed himself for the disasters that befell strangers at bus stops; he blamed himself . . .

. . . and Dinah left him to it. She anticipated the next cigarette she would smoke, because Dinah had taken up smoking.

She smoked cigarettes greedily, sensuously, down to the filters, sucked them empty with hollowed cheeks and covered their odour with Freda's eau de cologne spray. The finished ends were pushed through the metal bars of the bedhead and fell behind to the floor, where they remained until Gerald the porter, the cigarette magician in clogs, came to make them disappear.

Gerald was a large, well-spoken boy with pretty blond hair and something weak and dangerous in his eyes. He came by in the mornings, his head and right hand peeked in around the door, a look of surprise on his round face as a cigarette magically arrived between his fore and middle fingers. Dinah had never thought to smoke cigarettes before. The ripped drowning lungs of Danny's father had given her a horror of the damage cigarettes could do.

Her first cigarette was the difficult one. She had been sleeping. She woke to see it next to her, balanced acrobatically on its tip upon her beside table. There was a noise, a cough, she slowly turned her head and there was Gerald standing in the doorway. He entered the room. He shut the door. He came towards her. He produced a pink plastic lighter from behind her left ear. He smiled. So did she.

His manner was charming, a cocktail party host. She smiled again at the handsome stranger, allowed a hint of the coquette to play across her lips.

— I'm sorry, I'm being frightfully rude. My name is Gerald.

Dinah flexed her voice muscles. The sound of her own voice appalled her these days. It quavered, croaked, trembled, fluttered. She had been using her voice less and less and she couldn't remember if that was because the sound of it upset her so much or whether it was just because she had grown too old and run out of things to say.

— Di? Yes? No? Tina? No? I'm sorry love, it's my fault. Give it another go. Good. Easy does it. No rush. Yes. Dinah? Yes? Dinah. I'm very pleased to meet you. Are you a smoker, Dinah?

She shook her head. His company delighted her.

— It's never too late to learn.

That was true. She had never thought of it like that before.

— Would you like to try one? It'll make you cough at first, you'll have to get used to that. Perhaps you'll develop a taste for it, perhaps not. But then there'll be the most marvellous fun of doing something that isn't allowed. I'll tell you what. Why don't I start you off with this one? See how it goes. Or is there anything you'd prefer instead? A little something for your pain? A drink? A cocktail . . .? I don't want you to misunderstand me, Dinah. So awful to talk about money but I am a businessman. I always give one cigarette free but after that I have to charge for them, you know how these things work, you understand. But I'll tell you what I'll do. Special offer. You can have your first three fags for nothing. How do you like the sound of that? Here. I'll light this one for you, see how you get along with it. Take it slowly, don't breathe it in, just feel the smoke roll around your mouth and let it roll out again. Come on. Just try a little bit harder. Good. And again, draw on it, not too hard, slow down, you've got all the time in the world, just enough to let the smoke in, almost by itself, you like the taste? and out, good, good. Do you think you can hold it by yourself? Between these fingers, these ones, that's right, very good, hold it still, yes, and to the lips and draw it in and hold the cigarette away, good, yes, and taste, taste, and let the smoke roll away, good. Now you see there, we've got a long ash on the end of it, we don't want to make a mess do we? just push it behind, tap it against the bars, it'll fall down under the bed, very good, I'll clean it up later, now again, to the mouth, close the lips around it, pull on the cigarette, maybe a little harder this time, maybe let a little *little* smoke into your lungs, oh that's wonderful you didn't cough at all, that's marvellous, you were born to this, and away and breathe out, very nice, and again, good, and what do we do with the ash? Do you remember? That's right. That's perfect. Excellent. That's wonderful. And again.

She wanted to be his good pupil. She wanted to be the best pupil he ever had. She would not let him see how the smoke troubled her lungs, she was determined not to cough, and she didn't. And whenever the smoke poked against her eyes she refused to blink or cry. The first cigarette made her dizzy. Her throat was raw and her head felt as if it was spinning free from the safety of her neck. But she was able to hold on to the cigarette by herself. She stayed vigilant over the ash, she never allowed it to grow too long before knocking it away down through the bars of the bedhead. She smoked the cigarette all the way down to the filter and Gerald told her that if he hadn't known it was her first cigarette he would have thought she'd been smoking for years. He showed her how to hold the filter tip with one hand and wait a few moments and then dab the embers safely dry with fingers wettened by her tongue.

Gerald sprayed some of Freda's eau de cologne around her bed and promised to return the next day with another cigarette should she want it. He patted her hands and retrieved the butt from the floor and he made it disappear seemingly by putting it in his mouth but she knew he hadn't done that. He blew her a kiss before he left and she felt very tired, still quite dizzy, but elated too, and grateful that he hadn't asked her about her leg or her comfort; and even though smoking the cigarette had made her more aware than she would have chosen of the pains in her leg and the looseness of her thoughts, she was already looking forward to the second.

. . . And Leslie blamed himself and he drivelled on and finally he left. Soon Gerald would be here in his blue overalls and white clogs, with his cigarettes and his charm; in the meantime she turned to the photograph of Danny and Sam and she asked her watching husband if it was wrong for her to smoke. He warned her, gently, benevolently, with the ironic humour that was customary to him, about the risk of causing a fire and she promised she would be careful about that.

THEY MOVED DINAH into a shared room. A nurse held her hand uncomfortably hard and shouted shrill lies about the future as Gerald loaded her onto a wheelchair and Dinah knew by his wink that she wasn't meant to show familiarity so she did not. They rolled her along a dusty corridor defaced by pictures painted by children with learning difficulties, past a gallery of operating theatres, past queues of waiting rooms. Finally they unloaded her in a windowless room. Tipped her from the wheelchair on to a fresh bed which had an empty cabinet beside it and a television set above it, connected to the wall by a flexible iron arm.

Three other ladies were in the room. They were laid neatly on top of crisply made beds. All were watching their separate televisions, connected to them by ear plugs. None looked up to inspect the new arrival. The nurse, who was pretty and Irish and who possessed the most fascinating arrangement of freckles, arranged Dinah on top of her bed, jammed her leg inside its traction, joined her by the ear to her own TV set. And after the nurse and Gerald had piled Dinah's possessions on top of the cabinet and left, Dinah discovered the cigarette that Gerald had hidden in her hair.

Her blood was out of control, that's what she heard the doctors say, to nurses, to students, to each other, to Lawrence. They injected her with drugs to make her blood thin and other drugs to make it thick, and the thicked and thinned blood pulsed to her brain like sperm, and performed bright colourful things to her co-ordination and her memories. Things which were thoroughly enjoyable.

Above her was the wide, generous plain of television. Gay

colours, handsome teeth, sharp bursts of gunfire, kisses, domestic rage, domestic appliances, fast cars, powerful chests, beautiful scenery, and every now and again, but only when she wasn't expecting to see them, the figures of Koko, Yoshomoto and Daniel protected inside the screen. Always, the suety metal taste in her mouth, the monkey pain that was worst in her broken leg, the rashes on her bottom, her loyal right hand that still did what she told it to, the green fog that was stealing the rest of her body away from her. Her relations with the same speeches and the same silences that at first she tried to make separate sense of but then realised that things were nicer if she didn't make the effort. (It didn't matter what Leslie felt sorry about. It didn't matter that Lawrence wanted her shares in the company.) The prayers and grumbles of the ladies in her room, Gerald's demands, his cigarettes, her husband and her son and his son and their secrets. The problem of finding the perfect sentence to say at the end. This was her world.

— I miss my son.

She said this to Vivien late one afternoon, after the silence had become too painful for them both. She loathed her neighbours, TV dummies, tied to drips and their television sets, kept alive by vitamins, other people's blood, and the flickering images of soap operas and actors' tanned skins. Dinah switched on her own television by the remote control unit on the wall each morning after breakfast but so far she had mostly resisted the ear plug. She liked to watch the images in silence, on a tranquil white beach as everyone she had ever known lived in front of her inside a madman's dream of home.

— I miss him too, said Vivien.

Automatically, motherly, Vivien laid out her hand on the bed and Dinah clasped on to it, surrounded Vivien's dainty Nivea fingers with her own, taloned, cracked, puffy, liver-spotted, and big. In the former days of her vanity, Dinah used to dislike the size of her hands and her ear lobes. She would cover the lobes with pearl clips, squeeze her hands inside soft gloves when she went out in her best.

She still had those pearl clips, or at least she had until Gerald and the nurse had moved her into the shared room. She hadn't seen the earrings since then and that meant they were either safely inside the bedside cabinet or else they had gone astray in the move. Daniel had bought them for her, a gift to celebrate their first wedding anniversary, and she had worn them several years later when she had secretly followed her husband.

In the time of his decline Daniel had occupied himself with three activities. Once, or was it twice? Dinah, dressed in her best, tan kid gloves, black skirt suit, her pearl clips hiding her ear lobes, had followed him from their house on Stamford Hill. The 253 bus to Whitechapel Road, then down the moving stairs at Aldgate station, a brief wait on the platform before his black-suited little body found its solitary place in the second compartment from the front of the train. There he sat, she watched him — she didn't need to hide herself, he was not going to see her, he was not going to see anything — riding around in tube train circles, sometimes scribbling lines on scraps of paper fished from his pockets, more often just sitting quietly, staring at nothing, a flop of black hair over his eyes, a vaguely hopeful expression on his beautiful ugly face.

The days that he went into work he spent all his time in a private corner of the workshop floor. A gorilla named Ivan patiently followed his mysterious directions, and the two of them built their stupid unusable machines. The rest of the workers ignored them; even Sam, who never took no for an answer, had learned by then the pointlessness of talking to his brother. After Danny had died his machines had remained in that same corner. Ivan couldn't tell anyone what they were for. He was only a factory gorilla, placid temperament, hard-working muscles, no intellect. Sam had never shared his brother's gifts nor any of his confidences except for the single devastating one. Howard did not possess his father's genius for engineering, but many things, they say, skip a generation. Perhaps Simon . . . ? His eyes were the same. His equal passion for solitude had to be made to mean something. Five of them, with brass plate names, and not one of them called Dinah.

And the third task, which she had uncovered after his death, or rather which Sam had embarrassedly told her half the truth of, was to commit adultery with red-haired women in a rented room in Pimlico.

Vivien held tightly on to Dinah's good hand and Dinah let

her, and if Vivien remarked to herself on the whiff of smoke that surrounded her mother-in-law, a lingering scent of tobacco, covered only partly by the reek of pricey Christmas perfume, or if she noticed that there were fewer objects beside her on the bedside table, because their companions had been given away in grateful payment to Gerald, she didn't speak her observations out loud.

7

GERALD WAS DINAH'S friend. She knew he was because, unlike the others, he never claimed to be. He was, he said, a businessman. A businessman couldn't afford to give out cigarettes free, willy-nilly, as if they were flowers offered to women he loved. He had overheads, responsibilities. He was providing a luxury service and he expected his due return. Dinah smoked her first cigarette of the day after the breakfast things had been tidied away, her second after the early afternoon examination, her third before visiting hours were due to start, her fourth, the dangerous one, before sleep, when only one TV screen was still on, twinkling away for the insomniac who had no friends or visitors. Her roommates were securely tucked up in their TV worlds; they didn't notice Dinah's cigarette habit or if they did they gave no sign.

Time passed and Gerald told her she'd earned the right to be entrusted with a cigarette lighter. He fixed the flame to low and gave Dinah lessons in its use. It was pink and plastic and was the prettiest, most useful thing she had ever owned. He received a silver bracelet in return, fetched out from the tray of precious things she kept in the cupboard compartment of her bedside table. Her account, he liked to remind her, in his charming mock-severe way, must not be allowed to lapse.

Dinah Fischelis 7 April 1918 16a15 Waymead NW3
Esta Fisch Seta Fish Esher 4 March 1910010
Simon 13 May 1842
Freda 01 435–999?
Ruthie 34? 3? 6? Ruth
Daniel Levy 66 Dovember 1948
Lazarus Singe
Howard and Valery Levy, number 42 The Bedweigh,
Chumpstek, London NWT

It sometimes wasn't enough to smoke cigarettes and dream of Danny. There were times when Dinah could not ignore the thing she was becoming, when she had to look at it, in the cold light of five in the morning, when sleep was impossible, when the pain that nagged at her leg was the only promise of wholeness, when the magical prospect of Japanese boys and monkeys no longer had any lustre.

There was a grey-green fog that rolled up her body, stealing her limbs away. Her left hand was gone. Her right hand was erratic. The fog had almost reached her throat. She saw her death ahead of her and it looked like the monkey from before, except this one was big, an ape, with vulgar coarse features, not pretty at all. The ape was inviting her down a country lane that curved into the tree-lined downward path of a gentle hill. And at those five in the morning times she would try to pull back, or at least hold her position with bare cold feet on the slippery path, resist the silent population ahead of her, just there, she could see them waiting through the trees, not far ahead, so easy to continue on downhill, join them, there was

Esther and there was Danny not looking at her but glad to know she was coming, that odd combination of self-sufficiency and neediness he always had, and there were her parents and everyone who had ever loved her and died. To hold herself where she was, to try to grip her shivering feet to the path that was now coated with ice, resist the pull, it was impossible, so natural to let herself slide down the hill and join them all but still there was something inside her that was defiant: she would write things down on paper, her own name, her last address, and the one before that and the one before that; and the address of the house where Howard and Vivien and Ruthie and Simon lived, and Leslie's address and Lawrence's, and Muriel and Arthur and Sid her bridge companions, and everyone's birthdays. And usually she would get the information wrong. Simon she could do every time but 1982 became 1842 for no particular reason that she understood. Vivien became Valery. Hampstead Garden Suburb became Chumpstek. Death was naughty for It surely had no need for Howard's address or Daniel's birthday. And at these five in the morning times Dinah would feel inconsolably cold and sometimes she would cry and she would wish that Danny or Simon or Howard at least was there to hold on to her good hand and keep her true.

9

THEY WERE THE most important words in the world. They were the only words left.

Howard would return, maybe in a moment, maybe in a few years, it meant much the same thing. She knew her son at last. He was like his father after all, there was the same destructive wildness that swirled out of a purity that she hadn't been able to recognise the first time around. She would stroke his bald head and tell him the words and — release, a fluttering of wings, the walk downhill to the final consummation — her job would be done. They could do what they liked to her, she would endure. The words kept her alive.

She had spent hospital weeks, hospital months, a hospital lifetime, waiting for the world to become a sentence. Cigarettes, television images, pain, fear, relations talking in opaque bubbles, the fog, Gerald, a monkey and its enemy the ape, they all were like chattering static obscuring the one true message. But finally, in a moment of perfect silence when the room and the televisions were almost dark, the words were revealed to her. There they were, elegant and unambiguous and perfect, her legacy, drawn in spidery black characters on a white linen banner held up by two smiling Japanese boys. She read the words, nodding, yes, mouthing them, wanting to sing them, this was the truth, but then, to her horror, when she practised saying the sentence for the first time she discovered that the fog had stolen her tongue.

She simplified it. Gave herself an easier task to perform for efficiency's sake while hating herself for destroying the poetry of the revealed sentence. She broke it down to manageable parts that kept the meaning at the expense of the music. The *m*s were still awfully difficult so she made herself practise them, she hummed whenever she could remember to, and the witches stared at her, trying to cow her into silence, and Gerald laughed at her and stole from her until she had no more left to give but still he stole, and Lawrence tried to bully her and Leslie whined, and throughout it all she waited and she hummed.

IO

VIVIEN HELD HER hand. Leslie wept. Dinah laughed. Lawrence talked with the doctor. A porter who wasn't Gerald was using a mop on the wet floor. The nice freckled nurse passed through and was the only one who didn't lie with a smile.

— Look. See if you can get through to her. We can't. She's been smoking cigarettes. That's not allowed. She started a fire. That's dangerous. Her leg is healing. Norman's done a wonderful job with that. But she's suffered more strokes and her mind is in trouble. See if you can get it to focus on something. There are too many of you here. She needs some rest. All of you will have to go. Leave her be. That's all there is to say.

Screens had been put up around Dinah's bed as if the sight of her might tempt her three neighbours away from recovery. The floor was wet and still foamy from the fire extinguisher. The wooden bedside table was charred.

— A fire, whispered someone, perhaps Leslie, saying the word as if he wasn't entirely sure what it meant but knew it was something grand.

— Dinah. Cigarettes? But —

Gerald had been neglecting her, no smiles or flattery to cloak the bare facts of commerce, a handful of cigarettes scattered beneath her bed covers, the bedside cupboard door pulled open, his payment found and snatched away for ever, a cameo brooch, a pearl necklace, her silver locket which had belonged to her mother and contained a curl of somebody's hair. He didn't use his charm on her because he no longer needed to. The number of cigarettes increased each time, and the gaps between his appearances got bigger. She had learned to drop crumpled tissues through the bedhead behind her. They were cover for the dead cigarette ends that lay on the floor, like her, forlorn. Apart from the visits of Lawrence and Vivien and Leslie she had been abandoned.

— I don't wish to appear callous but a hospice might be the best place for her. She has been here a long time. There are limits to what we can do.

In the mornings she was lifted and emptied and dusted and made to lie in a neat line to watch her silly television. So she no longer squeezed the burning filters with salivery fingers. She dropped them behind her, still red, still glowing, reached for a tissue from the box with the hand that still sometimes obeyed her (the fog was moving up again, tickled her throat as it advanced; sometimes she tried to stay awake throughout the night, in order to see what the fog did to her body when she wasn't around to stop it), and she tossed the tissue overboard. White, virginal, and food for fire like Joan of Arc. And for the next hour at least she would thrill to her secret, while Vivien performed crosswords, while Leslie blamed himself for all the suffering in the world, while Lawrence connived at getting her to show him her will, to find out if her two per cent of the company would come his way — and she always feigned sleep during this time, bored by him and worried too that Gerald might pass by and learn that she owned more treasures than were contained in her bedside cupboard. Bright shiny licks of flame made familiar shapes in her imagination, it was all she could do not to giggle with the private glory of the thought of it.

— Here. Take it. You want more? Call my bank manager.

The reality when it came was a poor reflection. The first cigarette of that day had been the lucky one. Tossed smouldering behind her and tissues on top. A party of circus acrobats with the faces of her uncles performed graceless tricks on the screen. The three TV witches twitched their noses, turned their evil heads slowly towards her. The circus acrobats were suddenly Koko and Yoshomoto and she clapped her live hand against its dead opponent. There was a pleasant warmth beneath her, a light crackling sound like laughter but nothing more, no inferno, no cataclysm. Gerald was passing (fortunate for her, he said afterwards in his most sinister way). He ran in, handsome jowls shaking, a moment to examine the room and its possibilities, and then he ran out again, returned carrying a red bucket, two more porters behind him. He killed the fire with a sprinkling of sand from some far away beach and the other porters, in a rewarding flurry of activity, covered the floor with fire extinguisher foam.

— That isn't what I meant, Mister . . .?

— Charlie. Just call me Charlie. I don't give a fuck what you meant. Just take my money and show me where to sign and give her some good treatment, the best. You understand me?

She had made her protest and it had worked and she only wished the fire had been bigger and now she waited for punishment to come her way. She practised her humming and waited for her family to leave.

He would have to punish her for her carelessness. Gerald told her this with some of his old charm mingled with a terrifying severity. It was the only way to learn. For three days she was made to go without cigarettes entirely, and that was almost a relief. On the fourth day he returned with five cigarettes, along with the pretty pink lighter which he had confiscated and which she had dreamed about in the nights between, when her vigilance had waned and she had fallen into sleep. He raised the bedcovers, rolled the cigarettes between her legs, balanced the lighter upon her breast, and looked at her with love while he stroked her right hand. Then he pulled away, lifted up the photograph of Danny and Sam, inspected it, nodded, looked to her and back at it again.

She held her breath, followed him with her good right eye, not daring to think or to breathe. She was guilty and whatever punishment he decided on would be just. Gerald was stern as he tore the photograph into pieces. A monkey screamed. An ape beat its chest. Japanese boys lowered their heads in torment. Danny and Sam were ripped into rough pieces and tossed out of sight into her bedside cabinet next to an empty jewel box, which she had owned since she was sixteen years old and which, apart from her right hand and eye, was the only thing that she could call hers.

STARVED OF FAMILY *contact the photograph fragments lay
hidden, like Simon. Except, unlike Simon, they didn't want to
be hidden. There was something emanating from that hospital
bedside cabinet. Dinah felt it, but could not communicate what
she felt. If Charlie noticed it he did not respond; he would have
been more preoccupied with the way the future was being
moulded into its perfect shape, with the fluctuations of the
rhythm of his heart, with the unreachability of Howard and
the impatience of Mr Chatterjee and the sickness of Dinah
whom, despite everything, despite himself, Charlie loved. Ruthie
might have seen them, her eyes were quick enough, but after a
brief, wild, and watery affair with a porter who worked in
hydrotherapy, Ruthie didn't like to come to visit the hospital
any more.*

*It was Leslie who rescued the photograph, and not because
he was sensitively responding to the sparks of family spirit
emanating from the bedside cabinet. History is made by minor
fluctuations in bodily fluids. A shot of adrenaline to steady the
assassin's hand. A dribble of seratonin or RNA or testosterone
or blood. And in this case by the abundance of mucus dripping
from Leslie's nose. He had neglected to fold a handkerchief in
his suit jacket (since his mother's death small details of hygiene
tended to be neglected). Desperately he hunted for a tissue. He
let go of his aunt. His hands scrabbled around on top of the
cabinet. He knelt. He opened drawers, cupboard doors. Until
— and it shamed him to do it, he turned his back on Dinah to
perform the act, but still he was worried that her three neighbours would see him and think less of the family as a result —
he positioned his face as much inside the cabinet as he could*

manage and wiped his nose on the cuff of his shirt. It was while he was tucking the soiled cuff away inside the jacket sleeve that he saw the fragments of a photograph scattered around Dinah's jewel box.

12

IT WAS AN outrage, a murder mystery that he might never solve. The photograph pieces he discovered inside the bedside cupboard. Poor Dinah had been sleeping with her eyes open, maybe some indication of bedpan discomfort to be seen there, old people, he should know, require high maintenance. Half an hour he spent on knees that screamed searching for each fragment and he found the last of them wedged beside some damp cigarette butts. He woke Dinah to show her what he'd found and she said nothing, just made that humming noise of hers. A hospital porter breezed in, informed Leslie it wasn't his job to change bedpans, and breezed out again.

It was a crime and it was his job to rectify it. After establishing with an Irish girl that his aunt's needs would be attended to, Leslie left. He felt important. He drove back to the factory faster than he was accustomed to along a conveniently empty Westway. He went swiftly to his office. He locked his door and sat at his desk and emptied the fragments over his blotter and he went to work.

It was not like solving a jigsaw. That was the approach he tried first. He began at the corners, ignored any piece that did not have at least one perfectly straight edge, as he built from the outside in. There was no sky to help him. Or colours, everything was shades of grey. And the pieces didn't weld so easy. Crispy fragments with brittle edges that needed a butterfly collector's tweezers. If he wasn't so angry he'd have felt like crying.

He started again. He hunted for an important shape that continued across two or three pieces. A man. Two men. Workshop dungarees. The wall behind them. Was that a clock? The two light-grey blurs had to go somewhere in the middle. Try it that way and what do you get? Sam's face missing half of one cheek sitting on top of Danny's body and crushing his frail neck. He shifted it all around, gently now, respectful of the old photographic paper that could, if he were clumsy, become so quickly powder. An arm became a leg became the side of a table he'd never noticed before. Where were they standing anyway? Stupid that he didn't remember. He thought Danny went on the left and Sam on the right. Was it the right-hand border that cut that table in half or was it the side of Danny's leg? Where was the rest of Danny's hair? Was that a clock in the background? Or a stain? Move in the piece that was making Danny's leg look broken and it was definitely a clock. The time was eleven twenty-five. Not so fast. It was five to five. He'd given Sam three arms. The photograph had turned from rectangle into triangle. This was harder than any jigsaw. He had to use his memory for the picture they give you on the top of the box.

What had been the flash of an arm all this time turned out to be the missing side of Sam's face. There was no table. It was a tray or something that was fixed high on the wall on the left-hand side. The photo was coming together. Sam's eyes, his big chest, Danny's clever left hand, the curve of a knee, things found their place. Must not rush. This was intricate work. This sliver of shadow, snake-shaped, where did it go? Between the two men? Or was it part of the workshop wall? Yes, to the left, beside Danny, it filled the space and Leslie began to smile. Now he was left with two spots of blurry grey-white, the first plastic combs, unbreakable, try them this way, then that, both were wrong, turn them the right way up, press them into place, the photograph was complete. Healed.

Now the really difficult part. Leslie turned over each fragment so carefully, trying to keep everything in its place. It was an impossible task. Move the pieces too close together and they broke. The rough edges glided towards one another and if he tried too hard to make them fit they snapped and disintegrated, gone. He lost Danny's left knee that way and a small diamond out of Sam's dungarees. People had laughed at him for keeping a history scrapbook. It came in handy at times like this. He peeled the protective paper away from the sticky mounting card, tore it from his scrapbook (using a ruler as a guide) and gently, steady hands, lowered the card on top of the restored photograph, patted it firmly down with each elbow in turn.

He flicked it over, squared it on the plastic laminate table top. Some pieces overlapped. Others were too far away from each other. Danny's far-apart eyes were separated by a crescent moon of white from the page beneath. The brothers stood on different floors. Sam was a hunchback with a ripped-apart chest. The clock was good. You could set your watch by it.

Leslie did not know where the idea came from. While he had been at his restoration work his mind had been focused, undistractable. And ideas were not his strong suit. But it must have been there, slowly creeping up at him. Leslie sat back. Leslie smiled. Howard the Irresponsible. Howard the Wife Neglecter. Howard the Bad Son. Howard the Unforgiveable who had not attended Freda's funeral. (Freda who had been so good to him like a second mother almost.) Let Howard see what it was like to have a parent shown no respect. A photograph of his father ripped apart and desecrated. Perhaps in his irresponsibility Howard might still know how to suffer a little.

Leslie put the photograph inside an envelope. He wrote Howard's name on it in big characterless capital letters. And then, in the same print, just smaller, he wrote Care Of Katz at the top. Leslie whistled as he unlocked his office door and went down the stairs to the reception area. He even signed his name to enter the annual table-tennis competition (an event which he had managed to avoid since the age of thirty-four) as he watched Barbara lay the envelope on top of Katz's in-tray.

And off the photograph went, in fragments, in darkness. Surrounded by draft contracts and financial reckonings and company registrations, then — always in darkness, it needed the family touch to bring it into the light — a brief stop at Katz's tidy office, then, repackaged, redirected, off it went: postman's sacks and vans and sorting offices, and it took off and it flew, surrounded by postcards home and love letters and tax demands.

The photograph reached Howard the Irresponsible in a resort

town on a volcanic Greek island as he lay working on his suntan and contemplating whether it was time to go to New York, find a life there by being the first member of the family to reach Jack Levy's original destination. Howard was tanned and full-haired. He had stripped his life of everything that was superfluous, everything that was accidental, everything that was in place only because habit or obligation or thoughtlessness had put it there. Howard had performed this grand operation and now Howard was bored.

Childless Howard. Parentless Howard. Wifeless Howard (although for several weeks he had been wooing — with credit card and bilingual edition of modern Greek poetry — a local girl called Maria who liked to ride motorbikes). Jobless Howard.

The last time Howard had seen this photograph it had been intact. He had sealed it inside a company envelope, disguised his handwriting and delivered it, only half maliciously, to Charlie. Now it had come back to him, in an identical envelope with writing on it equally disguised. The photograph had been poorly rescued from some unimaginable act of violence. It looked like a maladroit child had mounted it on sticky card. Howard pushed aside the rest that Theo the postman had brought him — tax documents to sign and credit card summaries and a note from Katz and a letter from Vivien. He propped the photo up against the coffee pot. The rips in the surface made Sam look larger and Danny even narrower.

Howard stood up on his terrace overlooking the harbour. He performed the bending routine he had learned from the book of Canadian Air Force exercises left behind by a previous resident. Inside him, selves that had gone ignored in the sun began to exert themselves. Howard went inside. Howard packed his bags. His head was hairy, his body was fit; he was, he decided, free.

The arrival of the photograph did what Vivien's letters could not do, nor Charlie's investigations or Ruthie's prayers. It brought Howard home.

FIVE

DOMESTIC LIFE

If you fit in, you can stay.

Manson family dictum

I

SHOPPING, VIVIEN AND Howard, a supermarket rendezvous, the romance of produce, the plastic eros of packaging. She has a shopping list for guidance, he only an old photograph, tucked away, where he can touch it if he needs to, in his inside breast pocket. They wheel their trolley through fresh fruit and vegetables, greedy tourists promiscuously plucking plums from Spain, tomatoes from Jersey, lemons from Turkey, Caribbean bananas, Israeli avocados, English cucumbers and (organically grown) potatoes, French apples, Guatemalan sugar snap peas, and then a quick swivel, a sharp trolley turn, and it's king prawns from the Atlantic, a side of smoked salmon farmed in Scotland, and on: pasta from Italy, pesto sauce from Wales, bacon from Denmark, lamb from New Zealand, chocolates from Belgium, butter from Normandy, cheese from Switzerland. And yoghurts. The place is insane with yoghurts. Either side of one central aisle, refrigerated display cabinets filled with yoghurt pots in bewildering, unsettling profusion. Greek yoghurt, Greek-style yoghurt. Plain yoghurts, live yoghurts, yoghurts from sheep and goats and cows. Live Greek goat yoghurt. Live Greek sheep yoghurt. Live Greek-style goat *and* sheep yoghurt. Vanilla yoghurt with fruit at the top. Live plain yoghurt with fruit on the bottom. Yoghurt with whole grains. Babies' yoghurt, yoghurt that's light and yoghurt that's lite, children's yoghurt *With No Bits*. Thick and creamy yoghurt, low fat yoghurt and very low fat yoghurt and slimmers' virtually fat-free yoghurt (live and sheepy and lite), chocolate yoghurt, *custard-style* yoghurt.

— Do you think they make yoghurt-style custard too? How can they tell the difference?

She shakes her head like a benevolent teacher might, tugs at the front of the trolley, navigates him away, to Italian coffee and Indian tea, champagne from France, wine from Australia, vodka from Russia, beer from Mexico, German bread, American-style bagels, eggs from the ideal rural barn (picture on the top of brisk little hens gaily clucking their freedom). The world is flat, divided into aisles and sunlit by neon. Weak-moustached store managers in russet uniforms stroll and patrol and draw meaningful marks on clipboarded lists. Babies chortle in pushchairs and reach out for the world. Teenagers run trolley races slaloming around the tired Africans mopping the floors. Mothers yell at smeary mouthed infants. Men in lumberjack shirts and leathers make assignations by the cat food tins. A platoon of Down's syndrome old-young men stand guarding the lane of Feminine Hygiene. It is lovely.

And when they return the house appears empty, theirs again. Upstairs is Simon, noiseless. Outside in the back garden there is just an empty black blanket to show where Ruthie had lain to catch the last of the sun. They unpack the bags, the kitchen work surface piled colourfully high, splendid with universal goods.

— Where does this go?

The bag of apples.

— Where do you think? Wash them and put them in the fruit bowl.

Howard pulls open the bag, apples spill, roll, he catches them, juggles briefly — one drops, he picks it up, rubs it against his wife's thigh.

— You're being silly, come on, we don't have much time.

— And this?

Slab of Emmenthal cheese, slid down her arse, poked between her legs, she restrains a giggle.

— Howard!

— And this?

New World wine, Shiraz, the neck nudged to the groin of her jeans, the barrel of the bottle rolled so both her thighs can read the label, she wriggles away, he has her trapped, leaning, half sitting against the work surface island.

— And this? And this? And this?

212

Plums roll against her, a pat of butter digs again and again into her groin, she tries to get away, he doesn't let her, pushes her back, the cucumber.

— Come on. Stop. Please.

His wife. Middle-aged cowgirl in denim. Broad face, high cheekbones. A wild flurry of grey-black hair. Narrowed green eyes flustered, excited; wide red mouth open (pink edge of tongue naughtily protruding between chapped, vaselined lips). Two children and twenty-four years of marriage have failed in their conspiracy to ruin her body that once thought it was going to belong to a dancer. His wife.

His wife does nothing to prevent him unzipping and unclasping her trousers; the cucumber, curious, pokes against her white knickers.

— Howard! Please. They'll be here soon.

Down go her knickers. He peels the cellophane away from the cucumber, touches it against her, she gasps.

— It's cold.

He keeps on with it. One hand presses her thigh back to the island, the other holds the cucumber to her, teasing her with it, an informal introduction: harder, then away, then back again, between her lips, pushes in.

— Howard! That's for the salad!

Pushes in harder, withdraws.

— Oh.

And again, in. Deep. All the way, he almost loses it inside her, she jerks, bananas knock off, bruise to the floor, he pulls the vegetable half out, pushes it in again, lemons and plums roll a fertility dance around their feet, he is merciless. He snaps open her denim shirt, kisses the red flush between her breasts, her knees pulse wide to the side, her eyes blink shut, then blink open, and Howard and Vivien look down together, husband and wife, at the eroticised wonder of the cucumber.

— You've changed, she says, looking back at him from the sink where she scrubs potatoes clean and tosses them into a pan filled with water. You weren't like this before.

— Like what? he says, looking back at her from the cutlery drawer where he is gathering clumps of spoons and forks.

213

— Oh. I don't know.

She scrubs very hard at a potato. A curl of hair falls down over her eyes which she blows back to safety.

— Say it. Like what?

— Like, so, I don't know, it sounds silly, *sexual*. As if it's the only way you can think of connecting with me. Genitals. You weren't like this before.

— And? Do you prefer me this way, or not?

— I don't know, she says.

IT'S SUMMER IN the suburb. Birds sing, flowers are open, music plays loud from curtained windows, men in sports shirts and shorts and women in sleeveless floral dresses walk hand in hand comparing Spanish holiday sandals. Ruthie sunbathes naked in the garden, challenging Howard, or so it seems, to respond in some dangerous way to her body; her brother stays locked inside his room. Howard is about to resume his life as a company man — but this time he's vowed to do it right, to master all the details of commerce, to satisfy the dead and the dying, to prove his skill with balance sheets — and Vivien is throwing a party.

— It's not a party-party, she says as she issues the invitations over the phone, and not a drinks party, it's something in between.

The drawing room has been rearranged. Dips and cold cuts and sandwiches and cold pastas cover the surfaces, and on the mantelpiece is placed a cut-glass bowl filled with Freda's famous cucumber salad, which is their giggling secret.

The party is Vivien's inspiration, to welcome Howard home, to tell the world that things have found their place again. Vivien first suggested it late one night, their third night, after they had run out of uninteresting things to talk about and do. They would have had to talk about Simon or themselves, truly. She would have shyly asked him about the things that he had done while he was gone, and he would have told her some of them, and then she would either have cried or shouted him away, and he would have had to explain why he had run and he still couldn't do that and the conversation would probably have finished off the pair of them. So instead she had turned to him,

215

a glorious smile on her face, the thin top sheet around her body, and she had said,

— Why don't we throw a party?

It was the reverse of what he wanted. He wanted anonymity, a slow settling in. He was done with grand gestures. But planning the party brought them together in a way that nothing else had. He reassured her that no one would think it insulting to his mother to throw a party while Dinah was in hospital, telling her this even while he himself felt it was the most obscurely insulting act in the world. Daring her to die. Pretending she was already dead. Together they drew up lists of guests, and wrote their names in eyeliner pencil on the back of the television guide from the cable company. They planned the menu, taking long half-asleep detours to remember food they had shared in different places (*Do you remember . . .?* and *Was it in Paris that . . .?* and *Yes, I know, I thought I would die . . . !*). They chose the piano music that would be their party soundtrack, and then, over the next days, spent a happy time inventing acceptable melodramatic scenarios — glasses of wine thrown into faces, fist fights (he and Charlie), unlikely love affairs (Tessa and Leslie; cousin Bertha and smooth Andy Astaire from Consolidated). The party was, at least in its anticipation, something that belonged comfortably to them both.

Ruthie has been bribed to attend, and Charlie will be there, the first meeting since Howard's return, and other long-ago people like Tessa and Leslie, and business associates and people who used to be thought of as friends; and Howard's only intervention was to make sure that Miles Rodgers from across the road will be there as well. Howard invited him personally.

There had been references to Miles Rodgers in two of Vivien's letters, both in passing. Howard received the first when he was staying in Cornwall, forwarded on by Katz. He had not taken any notice of it at the time, read it several months later — to be satisfied by the levelness of all it described, and to be only mildly piqued at the mention of a handsome man.

Dear Howard,
How are you? We are fine. It is a beautiful sunny day,

216

perfect weather. I look out of the window as I write this and I see Miles Rodgers supervising the laying down of terracotta tiles in the patio outside his house. He is very fussy with the workmen who exchange murderous glances with each other when Miles's back is turned. The tiles are Italian and produce a very nice effect. The monkshood we planted is in full purple bloom.

The children are fine. Simon's play was an absolute triumph and he was definitely the star (this is not just a proud mother talking) even though all of them were wonderfully good — except for the boy who played the curate who kept forgetting his lines and going red-faced under his paint and his mumbling voice was quavering the whole time and it was quite embarrassing. Ruth came to the first performance too and laughed herself silly. It is such an endurance test, all those examinations, and she is so much *lighter* now that they're all over. She has developed a passion for physics which took me completely by surprise. I thought at first it was one of her 'interests' that she would show a wild enthusiasm for and then drop. But it seems as if I was wrong. She spends a lot of time at the Science Museum and at the library in the company of some very studious boys and girls from school, even though her exams are all over and she has earned the right to play. I am quite excited by this development. It seems as if she might have turned the corner at last.

Lizzie is deathly serious about the gallery-shop idea, except now it is a gallery-shop-restaurant-bar idea. We inspect the premises almost daily, in the company always of Lizzie's architect John, a young, very patient man, who is, apparently, flavour of the month in the architecture world, and who is also almost shockingly handsome. He has those Italianate looks that Lizzie has always been such a push-over for. But he's very respectful, not at all arrogant, doesn't put on airs, is full of fascinating ideas, and is entirely tolerant of Lizzie's daily changes of mind. (But then I suspect she is paying him a fortune — at least — for his expertise, tolerance and looks.)

You do remember Lizzie, don't you? She wears less make up than she used to — back in the old days with that paled face and whitened lips. Her blonde hair is now half-grey and that suits her greatly. I think it is her enthusiasm that keeps her so young. Ruthie is very impressed with my foray into the art world. She's patiently tried to explain to me the connection between John and some big names in the fashion industry but I get completely lost. I've had to promise her a job in the complex if it ever gets off the ground.

Simon is tremendously supportive and wise far beyond his years. He is very pleased for me that I'm developing an outside interest. He plays quite a lot of cricket and saturates himself with TV football from some championship somewhere.

All the best,
Vivien

It was a happy world she described for the two of them to believe in. Simon the sportsman and school hero. Ruth the scientist-to-be. Vivien, the gallery owner, reporting back domestic happiness to someone who was too craven or stupid to be part of it. It was a life improved by his absence, a life which, he sometimes had felt, justified his absence.

— Did you know I was reading them? Didn't it matter that I never wrote back?

— I knew they were getting to you. Katz has always been reliable. No, it was better that you never wrote back. It made you the perfect audience.

The letters arrived in batches. Katz would tend to let them build up, and then he'd send them all together. The only other letter to mention Miles Rodgers had been sent the following spring. It had been a romantic season.

Dear Howard,
Ruth is in love! It is wonderful to see, her good nature flowering, the flush on her face, the dancing hope in her eyes. Her eager lover is a boy called Ben who is studying for his science A-levels in preparation for medical school too. He is a very steady boy. Scholarly, careful, ambitious, clever (Jewish!), entirely smitten with our Ruthie and a very good influence on her. I had worried (I had hidden this worry) that she would go for the adolescent excitement of boys with motorbikes and drugs, that the dangerous streak in her would overcome all the good things, but her boyfriend is so removed from all that. (And he is very handsome too.) She is settled in love, we wait for Simon's turn now.

The house is recovered from the little emergency we had. I don't think I told you about it — a flood, like a natural disaster, and we the refugees hugging our most valuable possessions tight to our wet breasts. It started with a washing machine hose that, unnoticed, had begun to perish. A suspicious wet patch on the scullery floor, like a bloodstain in a ghost story. Each morning I would mop it up. Each evening it would be there again. It was hard not to get fanciful over the whole thing, a mark of guilt, domestic setting, horror. Ruthie (or Ruth I should say, she is very grown up now and suitably pompous with it and doesn't like diminutives) particularly was unsettled by it. She took it personally, and blamed me for it, and somehow, in some way, the stain represented you. I kept putting off getting a man in. I don't know why, probably pride I suppose, the idea that 'a man' would solve this for us was annoying to me. Simon's reaction was very similar to mine, a curious sort of inertia muddled with grim satisfaction.

Then the disaster. We'd all gone out together, a family pizza outing, and when we got back the whole kitchen was under water, *inches*, slopping against the garden door, soaking the hall carpet. I called the emergency plumber, and Ruthie ran across the road to fetch Miles Rodgers who wasn't in. She came back with a new neighbour, Mr Marks, a sad-eyed man with brown hair and a grey moustache. He was very shy and obviously hopeless at practical matters and he just looked very worried and awkward and then he went away again slowly and melancholically. The plumber came and quickly found the offending pipe and replaced it and went off whistling, leaving behind one of his monkey wrenches by mistake and an outlandish bill on purpose. Ruth was in high dudgeon after that. I couldn't tell you why. This for us is adventure.

The gallery is going well. We are changing exhibitions soon — the photographs stay but Lizzie is getting a new painter in, a decision which I feel is neither entirely artistic nor commercial. Her screaming spats with Edwige are becoming legendary. John has gone off to Kyoto to design his bank. We had a sad and sentimental parting over midnight doughnuts in an Arabian casino as we considered the things that might have been had things been different.

With love,
Vivien

It was the first of her letters to be signed *With love*. Before that it had been stern *I hope you are well*s and polite *All the best*s softening and thawing slowly at wintertime into *Warmest good wishes*.

— Where did Mr Marks come in? Did you make him up too?

— Sort of. No. I gave him the name. He moved in at number fifty-three, floated around looking melancholy for a while and then moved out again. I was thinking of having an affair with him.

— Oh really. Did he know about it?

— No. An affair, I mean, to write about. Not an *affair*-affair.

— Why did you do it?

She looks at her watch. Only the imminence of the party is allowing them to talk like this.

— They should be here soon. That sounds like a car.

He goes to the window, draws the curtain back to inspect the road. His other hand scratches at a Greek mosquito bite on his shoulder.

In her second letter she had been almost hysterical with rage and contempt:

> I am not going to give you bad news. You want to be told that your children are suffering, that they need their daddy, and your abandoned wife has turned into a dipsomaniac shoplifter without you to pleasure her into righteousness (there are things I could tell you about that! — why are men so stupidly cocksure?). Or is it that you want to be told chunks of normal things, the banal highlights of daily routines that will make you proud of your children and lustful of your wife so your self-pity will be fed and you will feel in your cups that you are entirely unnecessary to all our fortunes?

They had agreed silently on the answer to that one, even if their interpretations had differed. The worst thing about returning was what he had done to their preferred world. It was a world in which Simon attended his sister's seventeenth birthday party (*He just appeared, cool as a cucumber, looking very handsome. Life here has taken on a new lease. We do not miss you.*), in which Dinah held Hanukkah parties for the family, where everyone, even Charlie, attended and sang.

220

Vivien received Valentine's Day flowers and chocolates (heart-shaped box, covered in red silk moiré) from an anonymous admirer she claimed to believe was Howard. Ruth and her romance and her studies flourished like the flowers in the garden. (*A giant picture, a floral flag that can only be truly appreciated from above, I suppose. So who then do I plant it for?*)

Simon plays football for the school, he is a clean-complexioned athlete, a scholar, the champion of the weak, the envy of the strong. He begins, tentatively, a romance with a girl called Carina, half-Italian, half-Scottish, and Vivien breaks the couple apart, and then out of guilt she brings them back together again.

If only Howard had stayed away then his daughter would become a doctor, his son remain an all-round school hero, dashing Simon, and his mother would not be sick. He was killing his mother by remaining in the real, old world. No, what had Vivien said? Her last letter, the one that had arrived in the same mail bag as the repaired photograph, the one he had planned on them all reading aloud together, the family together, in laughter and scorn. It had, regretfully perhaps, told the truth about Dinah.

— I came to rely on those letters, says Howard.

He lets the curtain fall back. He is waiting for her to admit that she did too. In all, she does not assume the proper shame or embarrassment about the lies she had told. Rather, she seems proud. Defiant, sitting on the bed, forcing a curl of hair into place behind her ear where it stays for a moment and then springs back into the same place she had diverted it from, she says,

— I told the truth about the flowers.

And so she did, the progress of her garden, the only details he became accustomed to skip, sedums and magnolia and japonica, iris reticulata, monkshood, lilac, their brief glorious lives.

As the gate chimes sound, as Vivien leaves the bedroom, to inspect her unstable arrangement of hair one last time before taking the stairs to greet her first guests, Howard opens the letter. It's a brief one, which dawdles only a little over Simon's twin appointments as Junior Prefect and football team Captain, and Vivien's rapprochement with Mrs Campbell, the icy ogress mother of Simon's girlfriend Carina, which briefly puzzles over whether it is a good idea for Ruthie to plan on taking a year out between A-levels and medical school, before, tersely, it finishes with: *Dinah is in hospital. She has had a stroke. All my love, Vivien.*

3

IT WAS AS if it was fancy dress. No one had said anything about fancy dress. A theme party: Come As Someone Slightly Older. Howard touches his own unspoiled crown for consolation as he examines the heads of the men around him. They whisper and stare and then when they think he is noticing they guzzle down Australian wine and French champagne and Mexican beer and pick at plates of food that comes from everywhere. Their heads are obscenely bare, bellies hang over belts and waistbands. The women act out age more subtly. A rheumatic way of bending, an aching eternity to surrender a coat, a trick of walking on straight legs that makes them look like potatoes balancing on cocktail sticks.

There are people in the drawing room and people in the kitchen and people in the hall and people in the garden, strolling, drinking, admiring Vivien's lavish use of floral colour. Vivien quietly, pleasantly bustles. Drawing room, corridor, kitchen. Kitchen, corridor, drawing room (where a Schubert piano piece turns discreetly on the cd player). A sideways kiss, a plate of smoked salmon, the tactful filling up of a glass.

There are two sorts of occasions when Vivien looks free. In the brief, still moment when sex is about to give way to its aftermath, and when she's playing the part of hostess. Vivien learned her hostess skills early, on air force bases aimed at Khrushchev, watching her mother, the Lieutenant-Colonel's wife, hide all her unhappiness beneath a pretty painted smile. (The Lieutenant-Colonel was a military psychiatrist who would have, his widow used to inform Howard, risen to a higher rank had the two of them not been — her voice would then vamp to a disconsolate valiumed whisper — Jewish.) Vivien had learned from her, if nothing else, that handsome arrangements of food and generous amounts of alcohol work miracles in mixed company. She alarms Howard with her efficiency, with the ability he envies to become a function of her task.

Howard establishes his position in a corner of the drawing room, where the wall and the window provide some kind of protection. No one has approached him yet, he is too glamorous for that. They are talking about him; he hears his name harshly whispered, sees the sly fingers pointing his way. Is he The Man Who Returned or The Man Who Ran Away? He wears the suit that Vivien picked out for him, with a belt to keep the trousers up. The drawing room is filled. Leslie is demonstrating to squeamish, immodest Ruthie how they used to dance to big band swing in his day. Charlie is introducing his second wife to his current girlfriend. (His wife looks pleased. His girlfriend looks away.) Leslie performs some intricate captivating stuff with his feet and Ruthie escapes. Is that poor unwanted Bertha by the garden window? Katz and fat Mr Feldman, the company book-keeper whose first name no one knows, sit silently in companion armchairs. Charlie is flicking almonds into the air, catching them in his mouth and bragging about something. A group of men, a bunch of *guys*, joshing and shoving and pushing, surround Howard in a blue-suited semi-circle. They look at him like he is too grand to touch and they discuss taxation, direct and indirect, when he knows that all they want is to talk about sex and freedom and desertion. (He is a hero to these boys. What do the girls think? He is soon to find out.) He watches Vivien steer Miles Rodgers's fork safely away from Freda's famous cucumber salad.

Howard has never considered the existence of Miles Rodgers before. As far as he can remember they have only had one conversation, while watching a neighbourhood fire, when Rodgers patiently, rather excruciatingly, explained the geographical meaning of the d in his surname. It has, apparently, something to do with Scotland. Rodgers is not a handsome man. He looks a little like Abraham Lincoln, and has done what he can to encourage the resemblance by growing a beard to cover the bony jut of his chin. His face is hard and pale and his deep-set blue eyes look petulant, and he's balding and has gone in for combing the hair over from the side, something which even in the worst days of Howard's alopeciac vanity he was never tempted to do. Howard tries to imagine him naked, with penis erect, tries to conjugate him with his wife — missionary position? doggy style? certainly long slow measured strokes of bonily jutting penis. (Circumcised? Unlikely.)

— I expect you're feeling pretty pleased with yourself.

— I'm sorry?

It's Tessa, Charlie's one-time wife. Estranged or ex, Howard isn't sure about her legal or social status.

— Well I think you're hideous, monstrous, foul. What you've done is disgusting.

— I —

— There's no point trying to sweet soap me.

— Believe me —

— Vivvy's a saint to take you back. It's monstrous, worse than a Nazi, treating her, your children, the way you have.

Tessa is the only one who's ever called Vivien Vivvy. She's looking old now, and dressed to show it, imperfect peroxide hair, smudgy lipstick, deep deep lines around mouth and eyes: how can she believe in her own rage when she looks so unbearably ancient? How can any of them believe in themselves, in their jobs? He would like to tell Tessa she looks old. Finally he would be able to punish her for her relentless habit of always calling people by unliked diminutives of their names. There is nothing to stop him. He can do exactly as he pleases.

— You're looking old.

It does the trick. Charlie's ex truthed into silence. Howard can do whatever he pleases and it will work. He is the dynamic one here. No one can best him. *Next!*

Miles Rodgers is drinking white wine with girlish puckering sips and listening to something that panting Leslie is saying. Howard goes over to join them.

— Howard. I have to say you look very well.

Leslie says it like it's a crime, that looking well implies sinister artifice or deep conscienceless crimes of the soul.

— You too, Leslie, never better. (There is a limit to his truth-telling. Why pick on sad old Leslie?) And Miles. Good to see you here, how are you?

They shake hands, examine each other's faces for signs of knowledge and deceit.

— It's very nice to be invited.

— I was talking, says Leslie, about the Assyrians. But tell me. (Tears prick his eyes, his voice turns funereal.) How is Dinah? Her condition is a little improved maybe?

Dinah is of the past but stubbornly hasn't left to live there yet. Her body has gone mad. Her skin has turned jaundice-yellow as her liver disintegrates. Her lungs are drowning in pneumonic phlegm. She has a urinary tract infection that is a delight to her doctors. Howard spends five minutes each day by his mother's hospital bedside deadside, trying not to listen to the rattle in her lungs, always leaving when her hand performs its stupid swiping movement over her brow. She is in her final coma, and if he believed in God he would pray for it to end. She no longer can attempt to talk. She had been trying to say something about machines. Or, equally, it could have been margarine. It is a relief that her meaningless mumming of *m*s has been silenced by merciful blood clots in her brain.

— Not so good. It's — what can I say? Not good. We get on with things. How's the cucumber salad? Up to scratch? Miles, you must try some.

Leslie is acting scared of something, was he always that frightened? Howard is startled by his own certainty that Leslie can't bear being near him.

— Why don't I fetch some for Mr Rodgers? It would be no trouble.

They watch Leslie go on his party travels. He passes by Ruthie who is wearing too much makeup and too few clothes and is dallying on the arm of an unpleasantly handsome long-haired boy who is slightly familiar.

— Pretty girl.

— My daughter.

— Yes.

Got your eye on her now? Suburban Bluebeard or Browny-grey Chinbeard at least, the mother is not sufficient to appease your monstrous appetites so it's the daughter you turn to now is it? Creep. But Howard can see Ruthie's appeal. It shocks him to find his daughter sexy. But she is. She carries five or ten extra pounds, maybe more, but that doesn't matter yet, it's just a becoming, innocent exuberance of flesh. Howard is awkward with her, ashamed by his own reactions. He would love to sit down with her, plan a future for her; so far, he has been unable to do that. Tonight she wears the boxer's boots bought with her party-bribe money, fishnet tights raddled with holes, a short dark skirt, steel chains around her neck, a baggy black vest with showgirl tassels and gaping armholes through which you can see — Stop it! Keep attention fixed on Leslie, an enemy of arousal if ever there was one. There he is, almost at the mantelpiece, detumescent dandruff on his slope shoulders, thick spectacles through which to ogle Vivien, who is filling the glasses of Charlie and his new girlfriend, a blonde girl who carries boredom past the point of elegance. Leslie waits. He tries one direction, Charlie's expressive arms block him. He tries another. He's made it, he's through. He's at the mantelpiece, where the cucumber salad awaits. Beside him Mr Feldman struggles to escape from his armchair. Mr Feldman pushes hard on each arm and falls back again. He remains there, panting. Mr Feldman wipes his nose and his brow with a grey handkerchief fetched from the breast pocket of his jacket. What a swell affair this is.

— Nice music. You've been away I hear.

What does he say to that? You've been sleeping in my bed I hear? You're bald I hear? You've been fucking my wife I hear?

— You've had some tiles put in on your patio, I hear.

Why did he say that? He's never knowingly used the word patio before in his life. And anyway it's a detail from a Vivien letter and therefore belongs to a different, better world than this one. Miles Rodgers laughs, as if he has just heard something witty. Then he looks solemn and slowly closes his mouth. His long yellow teeth are disgusting.

— That's right. Awful expense and really I have to say I miss the lawn. Nice music. Is it Schubert?

Annoying that the sap got it right. Schubert will be ruined for him for ever.

— You know your music.

— Absolutely. But classical is a little off my beat. My specialities are Broadway show tunes and jazz before it got modern.

— You arsehole.

— Excuse me?

Howard hadn't meant to say anything rude, to come out waving his colours and firing his guns. But anyone who likes musicals is bad enough and on top of that jazz *before it got modern*.

— I'm sorry. It just came out. Nothing personal. (He allows his cuckolding neighbour a smile. It dazzles him.) Would you like something to eat?

— No. Not yet. I . . .

Howard had expected Vivien to have better taste. Unless this man is hiding a monster in his shorts he's just what he appears to be.

— You married, Miles?

— Divorced, I'm afraid.

Afraid? Afraid of what? Afraid of women or afraid of sex or of paying the bills or getting found out fucking another man's wife?

— Afraid of what?

— I'm sorry? Oh, I see. Nothing really, figure of speech.

Afraid of nothing, that's better. Miles Rodgers, don't forget the Scottish d in his name, suburban stud superhero, daredevil Miles, the man without fear.

— Cucumber salad for you Mr Rodgers. It was my late mother's special recipe.

Leslie saves the day. Howard was contemplating violence. A passion to hurt that he's never tasted so sharply before, except maybe on occasions directed at Charlie. Those long fascinating teeth crushed by his fist, shattering down the gullet. Those sad blue eyes popping with pain, the strand of light brown hair sliding from its place to hang uselessly down the side, dangling over Mr Rodgers's ear.

— Looks lovely.

Leslie delivers the bowl and looks desperately around for an excuse to depart in speed. Eager-to-please Mr Rodgers gulps down Freda's famous salad, slices of cucumber marinated in lemon juice and pepper and Vivien's uxorial vagina.

— Thank you. Umm. Delicious.

— Do you have a wife, Miles?

— Well, ha ha, I used to.

Vivien has been observing them for some time. She wants to come over, wants to break it up. Charlie saves her the trouble. One arm draped around Howard's shoulders, the other around Leslie's. The smell of brandy for everyone.

The party is a mistake. It is all so precarious, the safety of familiar places. He thought he could slot right in, factory cog, and in a way he has: his daughter calls him Daddy; his wife allows him intimacy with vegetables; but he can feel the edges of things (and on the other side, what?); and his body still doesn't fit inside his old clothes. Things will have to happen. He knows that life will change, somehow, perhaps unpredictably, when Dinah dies. Their last common link, a past that they all, through her at least, share. He has done most of his grieving for his mother's death in advance, sudden tears while shaving, or under his sunglasses while uncomfortably keeping his nude daughter company in the garden. It is down to the business. LPC will be the proving ground where he can bury his differences.

— You moulting, Howard? Fine head of hair all the same.

Charlie's first words to him since his return. Straight in. Finding the significant thing. There are hairs resting on his shoulders, but — a sly hand checks — no damage done to the crown. The effect of the season. Cats lose their fur, snakes their skins, Howard his superfluous hair.

— I'm coming back to work tomorrow.

He doesn't know what he expected. Something dramatic. Nobody's glass drops. Nobody's conversation abruptly stops. No gasps, no guilty conspiratorial glances. His enemy does not blanch. Charlie looks happy, Leslie looks mournful, but Leslie always looks mournful.

— That's grand, boy. We've got a lot to talk about, moving ahead on all fronts, let's save all that for Monday. Meanwhile, what do you say, let's get stoned.

Charlie lights up an extremely large joint. This was, is, his enemy. This is the man who — it had to be Charlie — lost him his job. Does he have to explore that now? Is that how to start? Is that the way in? Pick at the wounds of the past in order to heal the present? Charlie takes a few heavy puffs of his joint and we've almost lost him behind peaceful clouds of smoke.

He sucks in, one two three, hissing sounds, and any smoke that has strayed from his mouth disappears, back to Daddy, up his nose. He offers Howard a drag and Howard takes it, then escapes to the stereo and faithful Miles follows him there. The Schubert is finished so Howard replaces syphilitic Franz with the most modern jazz he can find.

The guests are getting drunk. Overripe women flirt with men who dream lasciviously about their babysitters. People look at Howard and try not to show they're gossiping about his return which, he is sure, they find even more shocking than his disappearance. Husbands envy him. Wives revile him. Bald men whisper the exotic word *toupee*. What does Rodgers want? Howard can't get rid of the loathsome thought that Miles wants to become his friend.

— What do you want, Miles?

— Just a top up please. Lovely wine.

Ruthie dances with Charlie. She dances very well. Leslie is staring at Vivien with pathetic greed. Howard has been away for nearly two years and this is his homecoming party and all he wants now is to remove all of his clothes and find a good place to cry.

— Tell me, ah, Howard, about your son.

— My son? Why? What?

— I'm interested. I was looking around the house, for the, ah, lav. Just a moment ago, a little while ago. I went upstairs, not looking to pry or anything, don't get me wrong — your downstairs was occupied. There was someone on the landing, Tessa? Marvellous eye for décor. We talked a little. I couldn't quite follow. Simon his name is? Isn't it? Mysterious. You do have a son, don't you? Where is he?

Howard sighs. He will give his usual response, picked from the moment when things first showed themselves to be wrong.

— Why is he called Porky?

Miles Rodgers considers or maybe just pretends to. This world is absurd. Strip away habits — the morning newspaper, the favourite lunchtime sandwich, the radio station you play in the car — and lose your memories and stop doing the things you do to please others, and what is left of the self? A few desires and then they peel off too, onion skins revealing nothing.

230

But Howard is different. Howard has the power to choose. Howard has returned in freedom.

— That's a joke isn't it? Porky? I don't know.
— Neither do I. Would you like to see the garden?

Howard manages to lose Miles Rodgers in the garden. Rodgers is boastfully glib with botanical names that sound like diseases to Howard. *Oh how marvellous, anorexia nervosa!* (Is that a dim Scottish roll in his throat or just a superfluity of phlegm?) He strokes leaves and fiddles with petals and coos and croons over alopecia and purple psoriasis and creeping hysteria, and Howard leaves him bent spryly double with his nose in a border — *Look here! Isn't this the most outstanding devil's knob?*

His love-rival behind him, Howard walks back toward the house and up the garden path. There ought to be a banner hanging from the roof — WELCOME HOME HOWARD! or SON! or HUBBY! or DADDY! There is no banner hanging from the roof. Simon ought to be here, rushing into his arms, holding his hand, walking beside him in unconscious imitation. Howard goes into the conservatory, scattering smiles and thumbs-up gestures as he goes.

— Seen your son yet?
— What?

What is this about his son? His son is, used to be, their secret. Not the stuff of party-party/drinks party chat. That long-haired boy is by himself in the conservatory. Sitting cross-legged in baggy jeans on a wicker chair, a glass of orange juice in his hand, Walkman headset around his neck. There used to be a logo printed on his black T-shirt for the world to read but it's been washed too many times for that. Is that Wiener Charlie is talking to in the kitchen? What would they have in common except some dead men from the unplastic past?

— Do I know you?
— You gave me a job once. Interview. Questionnaire. In depth. Fucking my soul more like. Chat. Welcome aboard, lad, you're hired. The responsibility of a fork-lift truck.
— Yes, I think I — you were a friend of Ruthie's.

He remembers the boy as a zealot of some kind. (Happy days. The Howard Levy Psychological Profile Questionnaire.

13. Are you superstitious? 20. Do you consider yourself to be a patriot? 15. What frightens you more than anything? 8. What is your favourite taste? 31. Do you embarrass easily? 32. *What are you thinking of now?*) The questionnaire had revealed something bad about the boy but Howard can't remember what and anyway he was always going to hire him because Ruthie had asked him to. The boy (name of? Paul? Matthew? Luke?) has a hard bony face, little expression, maybe the slightest suggestion of cruelty. He holds himself as if he is protecting something precious. One of his daughter's undesirable boy-friends. Not undesirable, silly word, desired certainly by Ruthie. Ruthie has always had an infinite capacity for desire.

— I used to drive around the company yard. Pick things up, put them down again. Nice party this.

— Thank you. Can I get you a drink?

— Don't drink. Fucks you up.

— I always thought that was rather the point of it?

— My boss. He loves to get fucked up. You name it, he'll fuck himself up on it. I tell him to keep a clear head. Stay on top of things.

— I'm sure that's very good advice. Look. I —

— I know. Chit chat. Reunions. Excuse my loquacity. Should auld acquaintance be — Seen your son yet?

— No. I haven't. But what—?

— That's all I wanted to know. Good. Sorry to keep you.

A gentle smile trespasses on to the boy's face. He takes a small sip of his orange juice and inclines his head to indicate their conversation is over.

Howard walks back into the house, feeling a little foolish and far from grown-up. The party is disintegrating. Ruthie has discarded the boots she bought with her party-bribe money and is dancing closed-eyed to her own inexplicable choice of music in the centre of the drawing room, the leather tassels of her punk-showgirl top performing their own, separate dance. Leslie sits spread-eagled in a straight-back chair rearing up on its hind legs. His moustache is dyed red from wine, his eyes stare at the ceiling. Mr Feldman, the company book-keeper, is attempting again to lever his massive body out of the deep-cushioned trap of his armchair, pretending with a convincingly happy smile after each failure that he wasn't trying to get up at all. Paintings hang askew on the walls. Married couples, enjoying their party divorces, pass their hands over the nearest exposed flesh of other people's spouses. Where is Vivien? Party goddess, it's unlike her to let things slip to this. Russell Parker, whose friendship Howard once, in a previous life, counted on, turns his back when he sees Howard look towards him. In a blue linen suit Andy Astaire from Consolidated (whose name, Howard, in another life, gratefully borrowed) is hoping to interest grave Harry Wiener in something tricksy that has to do with money markets. Wiener is like a big old daddy buck elephant. Grey wrinkled skin, eyes that have seen too much, reassuring bulk that is comfortably carried if oddly distributed, it would be so easy just to shut eyes like Ruthie and fold up inside.

Charlie is holding court in the kitchen. Swigging from a brandy bottle and smoking a big joint and getting people to make themselves ridiculous by appearing to be so himself. Charlie is the enemy. He always has been. And if it were not for the reassuring odious familiarity of him, there would be no place for Howard in any of this.

— The boy Howard! Rogue! My partner! Come to Charlie!

Pretend not to hear. The stairs are quiet. Peaceful. He wishes his mother were here. The first-floor landing is quiet. Not so. Vivien's voice drifting down. Howard stops on the stairs. She's talking. Pitched low. Her rich voice. The first time Howard heard Vivien speak on the telephone he knew he was going to marry her. He still remembers, cherishes, the first word she uttered in their first telephone conversation. It was *Oh* and it was said softly, almost bashfully, and it contained an eternity of meanings.

— . . . I don't know. Really rather silly to pretend that everything is normal, how can it be? He drifts around the whole time smiling, he never smiled like that before, won't tell me anything, just smiles, like a madman with jetlag. We keep missing each other except when — I don't know. Can I tell you about that? . . .

Who is she talking to? The tone is intimate. Confessional. Her lover? It has to be, but it is so brazen, Vivien spilling the secrets of her heart to creepy Miles Rodgers. Howard walks slowly up the next flight of stairs. He can picture the scene: Rodgers smiling encouragement, mouth open just enough to reveal those ugly yellow teeth; wife and lover would be sitting on the edge of the bed maybe, thrilled, sweating at the delicious open-doored illicit indecency of it, and now Rodgers's hand comes slowly out, Mr Reassuring, he's earned himself a grope . . .

Howard reaches the second-floor landing and sits with his back to the banister post. He stares at an old coffee stain on the wall, at the scuff marks on the skirting board. They're not in the bedroom. They're towards the far end of the landing — in the open, by the children's bedroom doors. Howard does not yet dare to look around; it's as if he has found his way into a Western movie and Vivien's is the voice of the dangerous gunslinger who's just sneered into town.

— . . . We were talking before, your, him and me. Before the guests came. Talked about the letters I wrote. He used to be so handsome, you know? I used to come to watch his group perform at places. Girlfriends can be so wretchedly devoted? Electric guitar with stickers and paint, thin suit, the one everyone had then, open necked shirt, standing so perfectly still — the girls liked him like that, so so hip: only I knew that he was being crippled by stage fright. A fringe of hair flicking into his eyes. How do we become what we are?

Why is she talking like this? What's Rodgers doing now? Nodding, sure, that's it, understanding in his eyes, monster cock tearing at his suit trousers like the Incredible Hulk.

— I wish. I know what he thinks. I didn't write them like a spell to bring him back, I didn't. Maybe I thought — silly, embarrassing, like sympathetic magic — that all I had to do was write these things down, and write them with love, and they would all come true. . . . I wonder how that happened. No. Excuse me, I was thinking out loud. And then it got too much. I couldn't. Not any longer. I wrote him a letter that contained all the truth. I told him about *everything* — you, and Ruthie in trouble with drugs and boys and about Charlie being up to something with the company. And about Leslie and his attentions. And I told him about my night with Miles Rodgers —

Who then is she talking to? To Tessa? Covenly secrets of one-time sisters-in-law. Or she's on the phone to her friend Lizzie who's her partner in the gallery. Who's her partner in the gallery that doesn't exist, hard sometimes to remember which are the true things.

— which I hadn't told anyone about, not even you. How does that make you feel?

Her voice drops, a seductress's stage whisper, and then a pause, and she starts to speak briskly, as if she has more words than time to speak them in.

— Which is worse? That I committed tedious adultery or that no one told you about it? I couldn't send the letter. I crumpled it up and I threw it away and the next day I spent an age trying to find it, and I did, carrot peelings, wet tea bags, and I tried to press it flat and now it's inside volume three of the Children's Brittanica which we bought for you all that time ago. Then I wrote the second version, the one that I sent, the usual lies, and then it was about Dinah, the state of her, it had to be said, it would have been a crime to keep it from him, his mother, and he came back soon after that even though he says he never read it . . . And now. We're almost like a family again. I wish everyone would go home. I'm glad they're here. It's nice and everything that they came but all the same. It's dreadful to say it, but —

Her voice drops even lower, below the compass of Howard's hearing. He creeps around the banister post, half-walking, half-crawling, like a sprinter trying to sneak himself a head-start, and there is Vivien, alone, sitting comfortably against the corridor wall, her hands resting on her skirted knees, her mouth at the level of the keyhole in Simon's locked door through which she is whispering her secrets.

— . . . maybe not, maybe I'm just not trying hard enough. When — what's that?

Clumsy crawling Howard. His shoe half-catches inside a banister strut, a quick thoughtless tug away, leather against wood — *squeak!* — his secret-telling wife twists her alarmed face to see him approaching.

— Howard! *Oh*.

— I didn't mean to startle you. Sorry. I was just — what are you doing?

— Talking. Nothing. Talking.

It would be silly to get up now. He crawls towards her, pats her knees, then slides away to sit like her, his back against the opposite wall. Between them a few inches of carpet and air. Next to them both, on his left, on her right, the wood of Simon's door.

She makes as if to stand up.

— The party. I really should.

— It'll look after itself. People are leaving.

— It's rude. I should see them out. I should —

— What were you doing?

— Nothing. Talking.

She doesn't look him in the eyes. They are both feeling guilty, as if he has caught her out masturbating after sex.

— Is he in there?

There are no sounds behind Simon's door. Why then are they whispering?

— Do you do this a lot?

She nods. Her eyes are upon him now, her prettily wide mouth is open. She's waiting for him to do something important, he doesn't know what.

— Every day. Yes. We all do.

— We?

— All of us. Everyone. He, Simon, he, likes us to.

Perhaps there are sounds behind Simon's door, listen closely: bare feet on carpeted floor, a hand feeling inside a pocket, a nose lightly blown, an adolescent boy's bottom settling down on the cushioned comfort of a bed. Somewhere a radio softly plays. What is the boy feeling? Hard to imagine. Pique perhaps, that Vivien's monologue has been interrupted.

— How do you know he likes us — sorry — you, to?

— I don't know. We just do.

— Does he ever talk back?

Vivien dismisses the question with an impatient shake of her head.

— He'd like to hear from you too. I know he would. He misses you.

Party sounds rise up from downstairs. Glasses tinkle, break. The music lifts up suddenly louder then cuts out, then loud again, a different tune. Voices are raised as if a fight is about to break out.

— Forget about them, Vivien says, coaxingly. Talk.

— How can I talk to a door?

She reaches forward, rests a hand on his shin.

— Just talk.

— About what?

— Talk.

He has never known her so firm. (Except perhaps once, when the children were small and Howard had suggested a puppy.) Howard shrugs. Tries not to feel ridiculous. If this is what is required of him then this is what he will do.

— Simon? he begins. It's your father . . .

Is he imagining the sounds? Light footsteps approaching the door on the other side. Is there someone there? settling down to sit by the keyhole? Howard clears his throat and leans forward and talks, awkwardly, as if he has to translate his words into English before he can speak them.

— How are you? It's been a long time.

He has to do better than this. Try not to think too hard, just talk. Vivien's fingers clasp harder onto his leg.

— We're having a party down below. Ah. Everyone's there, it's a sort of celebration. . . . (Howard looks to his wife for encouragement and receives it) . . . Charlie's getting drunk, and Mr Feldman's stuck in a chair and your sister is dancing. Leslie has had too much to drink. It's strange to be back. I've, we, we planned this party, your mother and me, it's a sort of welcome home affair . . . This is ridiculous. I can't . . .

— Go on. You're doing so well. Both of you.

— Ruthie wears too much makeup and I'm not sure about her clothes. Maybe you'll find this out if you ever have children but there's something about being a father and seeing your daughter through the eyes of other men. She's changed a lot. I wish I knew what you looked like. Do you shave? Is your hair long? Do you cut it yourself? Your mother is still very beautiful. It, alarms me.

It is strange, impossible, the experience of talking to the

door. He wishes Vivien were not there, her presence inhibits the secrets he would like to tell. Vivien was right, Simon's interest is tangible, the greedy, ideal listener. Howard would like to tell his son about some of the things he did while he was away, and he would like to tell his son about the difficulties of being back. Why doesn't Vivien realise that her presence gets in the way of this?

The noises below increase. He had been, for a moment, oblivious. Loud voices. Trouble.

— Look. I don't know what's going on below. I might have to sort it out. I'll be back later. Talk to you then.

It is a wrench to leave but also a wonderful relief. Husband and wife rise, hand in hand, he leads her down flights of stairs. The hallway is empty, the front door open. Outside, by moonlight, the last few guests present sombre backs in a line. Howard pushes through. To see Andy Astaire from Consolidated facing up to the long-haired boy. Astaire is performing some kung fu-kick boxing fantasia. The boy is standing almost perfectly still, poised, weight shifting up and down on the balls of his feet. He's smiling. Beside them — the cause of this dispute? — Ruthie is hugging herself and laughing. The long-haired boy has had enough. He moves forward to Astaire and there is something in the way he approaches or maybe it's something revealed in his eyes or maybe something said too softly for anyone else to hear that makes Astaire abruptly stop his strutting and take a step back. Howard is about to intervene when Charlie beats him to it. The boy allows Charlie to wrap a fatherly arm around his shoulder. They walk slowly off, watched by Astaire who calls out to them when they are nearly gone, *You don't frighten me!*

Charlie and the boy disappear in darkness. Up above, Simon's bedroom light shining purple through his curtains and open window provides a momentary distraction from the disappointment of thwarted violence. Fingers point, voices whisper. The legend of The Boy In The Bedroom that has been told too many times is told again.

Howard follows his daughter back into the house. In the drawing room Mr Feldman is still in his chair. Mr Feldman would like to go to the toilet. He pushes up out of the chair, rises an inch, maybe two, falls back again. He tries a second time, doesn't even make the inch this time. Mr Feldman sighs. Mr Feldman lifts his handkerchief up to his eyes.

Howard follows his daughter into the kitchen. She giggles. The other guests are gone. Charlie has piped them away. Glasses are broken in the sink. There is a wet patch on the floor that Howard nearly steps on.

— What was that, it about?

Ruthie doesn't say anything, just smiles, pours herself a glass of water from the refrigerator. Howard returns to the drawing room.

Mr Feldman rises, three inches, four, five. His hands grip the sides of the chair, he pushes up from his hips with all of his strength. He falls back into place again. He whimpers for help. It takes four of them — Howard and Ruthie on one arm, Vivien and Leslie on the other — to tug Mr Feldman out from his chair. He tells them as they heave what the doctor said about the state of his heart. He puffs out his cheeks to help them. His nose is sweating. Finally he stands. Vivien passes Mr Feldman his hat and coat. Howard sees him to the door. The party is over.

4

HOWARD LEVY, AN executive reborn, walks stiffly up the stairs in his old business suit. The suit is navy blue, pinstriped with red, expertly taken in at the waist and let out at the shoulders by seamstress Vivien. The black attaché case he carries is empty apart from the photograph of Danny and Sam. He couldn't think of anything else to put in it. The company tie he wears — yellow background, red carbon atoms and blue hydrogen atoms bonded together in Leviex precision — chokes his throat.

Up the stairs to the second-floor landing, Howard follows the worn blue carpet trail to Simon's door. The breakfast tray is ready for collection. Cereal bowl flaked with wheatgerm, ringed by milk. Toast crumbs on side plate. Knife and spoon licked clean. Racing car mug with coffee dregs at the bottom. Four empty bottles of beer from the previous night, Simon's way of joining in the party. Howard picks up the tray, crockery and glass slide over alpine mountains. On the other side of the door comes an answering sound: a muffled hand brushes against the wall, a shoe squeaks.

Howard replaces the tray on the carpet. On the other side of the door something electrical is switched on and off. The things this boy must know, if it is true that they all come to him. That must include even Charlie. Howard taps upon the door. Silence. But not the silence of absence, it's the silence of anticipation, breath-strainingly held, hands clutched in anxious fists. Howard walks away: on the other side of the door, his son approaches. Howard walks back towards the door: his son walks away.

— I'll be late for work.

It is absurd to talk to a locked door, therefore Howard pretends he is talking to himself. Even so his voice sounds awful to his own ears, thin and raised and shrill. There is a cushion he hasn't noticed before, orange and green African weave thoughtfully placed in the corner by the door jamb. Howard sits upon it, his back against the corridor wall, his head forward, his mouth towards the keyhole. He clears his throat. On the other side of the door his son slides down to sit as well.

— Simon. Simon? It's your father.

Howard shifts on the cushion. There is a pain in the small of his back and he's holding his shoulders high and awkward and there is no sign that anyone is listening to what he's saying.

— I'll be visiting your grandma later at lunchtime. It's the routine. Call the hospital at nine, after breakfast, telephone rings and rings until someone deigns to pick it up. Some under-trained overworked girl with a jaunty voice — Mr Levy? Of course. No change — Then I visit her in the afternoon. And there is, will be, a change, always is. Skin yellower. Eyes deader. More tubes sticking out of her. What can I do? Stuff a pillow over her face and press down hard?

His treacherous words produce a reaction. A gasp, pause, then the slow exhalation. Howard is tempted to go on in this vein, to shock his son into proving he's alive.

— Maybe there is something going on inside there. Maybe despite what they say life of a kind goes on. Of course it doesn't. She's dead and the body lingers on. Call again in the evening, after dinner (all my contact with her is ruled by mealtimes, but that's a Jewish mother for you). — Mr Levy? Of course. No change. — It's amazing, you'll learn this one day, perhaps you have already, no matter how strange things are, you go away, you make your own place where everything is new, and no matter what, you end up in a routine to keep things ordinary.

Activity on the other side of the door. Mental fidgeting. It communicates. Hermits get bored so easily.

— All right. I'm not going to stay here philosophising all day. I've got things to do, you know. Work. The company. Do you know how hard that's going to be? I don't think anyone does. I'm sorry. I didn't come here to whine at you.

They sit in silence. Howard is sure that his son is mimicking his own seating position. Perhaps he has a twin African cushion

241

to perch upon. Howard tries to picture his son. A fifteen-year-old with an empty face. A twelve-year-old scarred by acne and sullenness. A nine-year-old with his face squelched in pizza (*Why is he called Porky . . .?*). A new-born baby covered in lanugo and vernix wailing its horror at being pulled out into the world. Is that the point of this? Womb nostalgia? Like Danny on the Inner Circle, finding a safe approximate place? Tempt him out with gynaecology, they never tried that, slip some wide-open beaver shots beneath the door, the immodest maternal vulva, maybe that would do the thing, like forceps: out he'd come, wailing, covered in hair and grease.

— Have you seen this photograph?

Not quite the same thing. Danny and Sam ripped and half repaired. His talisman. He keeps it with him always. Howard flicks open the attaché case (H.A.L. monogrammed in gold beneath the leather handle), takes out the photo and starts to slide it through, ignoring the momentary panic that always hits him when he is without it.

— It's what brought me back. Sort of. Your grandfather. My father and his brother. They started the thing. I don't know who sent it to me. Katz passed it on. I wanted, before it's too late, to — this might sound hypocritical — to do the right thing.

Four hundred and sixty five thousand three hundred and nine. He likes to listen to late-night phone-in programmes. Peter from Putney with his electric piano, playing old favourites from the musical shows; Mildred from Edgware and her endless pogrom from the gas board; Eileen from Maida Vale, an antique American actress, strung out on drink and pills and lack of sleep and her memories of when she was frankly irresistible; hearty Harry from Hackney with his undimmed hopes for world communism. Simon likes their voices, their hopeful, slightly suspicious offering of night-cracked theories, he likes the bubbling urgency of the things they don't know how to say, and he likes the distance they keep from him.

Four hundred and sixty five thousand three hundred and ten. This is his world and it measures fourteen feet by ten. His world has a door at one end (patched with irregular ovals where the animal stickers have been impatiently removed) and it has a window at the other, barricaded with a thin purple curtain. In the corner next to the window is a sink where he washes and shaves and trims his finger and toe nails and sometimes urinates. Beneath the sink is a chemical toilet. His world is walled with paper that pretends to be pine wood. Against one long wall is his writing table, against the other is a long bookshelf and beneath it his bed. The chair with its pink-cushioned seat performs double service — during the day it is by the table; at night it is at the end of the bed, an improvised rest (the capacity for invention runs in the family) for his feet that long ago outstretched the child's mattress. In the centre of the ceiling a bulbous green glass lampshade. Across the ceiling stars and suns glint in the dark. His bed is covered by a racing driver quilt. His transistor radio is usually on the window ledge. His wardrobe is filled with books. He keeps his clothes in the cupboard behind the head of the bed. There are notes here too, and some letters, and his discontinued Dream Diary in a 1989 appointments book. The window overlooks the street. The window is tall enough, when fully open, for a large boy to climb through.

Four hundred and sixty five thousand three hundred and eleven. Simon wants to visit his grandmother in hospital. He would like to go there. He knows it would help her. His mother is there now, his father will be there at the next mealtime. Simon had considered going. He had got ready to go many times. He had put on his jean jacket and anorak. He had laced up his boots. He had slotted a tape inside his Walkman. He had wiped down his apparatus. He had unlocked his bedroom door. He had taken the steps down to the front hall. He had even stood at the front door. He then returned to his bedroom. He stacked his bowl and plate on top of the tray and left it outside the door. He then locked the door. It was too difficult to go. Simon can imagine her there. He pictures his grandmother's fear and shivers at his picture. He sniffs the sweaty formaldehyde air and nearly chokes on the taste of it. He hears a woman's shrill voice praying for help in a nearby room. He scratches his leg, itchy under its imagined plaster cast. He opens his eyes and stares at the make-believe wood grain of his bedroom wallpaper until he remembers what it is. Four hundred and sixty five thousand three hundred and twelve.

Four hundred and sixty five thousand three hundred and thirteen. Simon closes his eyes again, he can almost hear his grandmother's voice but there is too much interference in the way, a radio, a TV, maybe three TVs.

Four hundred and sixty five thousand three hundred and fourteen. He no longer remembers when they first came to tell him secrets. It might have been his mother, furious and abandoned and hungry, who whispered to him the temporary truth that she hoped his father would never come back to them. It might have been his sister, in between insults, telling him the sexual secrets that she couldn't afford to confide to her friends. It might even have been his father, telling him on a sweleringly hot spring morning that he would take him to a football match the next day if he came out of his room. But his father had disappeared later that day and anyway Simon thinks that the confessions started after that. He has never asked for secrets. He has given no sign of wanting sexual fantasies and sexual exploits, business treacheries and ecstasies and the exhilaration of driving stolen cars. He never thought that he needed to know the histories of other people's victories and defeats. But he has come to rely on them.

Four hundred and sixty five thousand three hundred and fifteen. At night, when the moon is narrow, sometimes out goes Simon on an excursion. He wears an anorak that is too small for him zipped up to the throat. A pair of baggy jeans and a further layer, of blue and white striped interlock thermal underwear, beneath. Black balaclava helmet around his head (a soft warming hug, allowing a few strands of ginger hair to peep up over the top of his forehead) and dark blue mittens on his hands. His eyes are almost bare, protected only by a pair of sunglasses; but his nose and mouth are hidden beneath the apparatus which he has invented himself, looks like a home-made branch of pussy willow — a tartan woollen scarf torn into strips and reinforced with cotton wool buds and sewn around a frame of clothing hanger metal; it presses cotton wool into his nostrils and urges its safe taste against his damp mouth, just enough room to breathe around. His ears are doubly pro-

tected by the fluffy purple ear muffs hooked over the top of the balaclava.

He crawls through his window, over the garage roof, slides down the automatic door, a quick look around, guerilla alertness, he checks his apparatus is covering as much of his senses as he can afford, and off he goes, down the garden path, the uncomfortable climb over the gate, he drops onto the pavement, and walks. He walks along the streets that lead him up to the Heath, trying not to look or hear or feel anything. Even a late-night light in an insomniac's window will fill him with solitary horror. He will taste the gin martinis on her tongue, shiver with her by a gas fire, be distracted with her by the phone-in station (maybe even call it up, put herself up for one of their quizzes, Hilda from Hampstead?), or listen to an old dance song and stroke her cat companion and slowly tear apart dead theatrical scrapbooks. The dangerous antidote to that is in the house across the road. The shared ecstasy of a couple's lovemaking twists his limbs, brings the blood to his face, pulls him jerking like a puppet into the middle of the road until the horn and headlights of a speeding car smash the spell and send him running to safety. The Heath is safe. Old and big and wooded enough to resist the human world. His customary bench. He sits on the back of it, rests his feet on the seat, loosens his apparatus and feels the chill wind burn his face and impose nothing upon him.

Four hundred and sixty five thousand three hundred and sixteen. His grandmother is preoccupied with machines. His father obsessed with his own father. Simon never knew the legendary grandfather. He looks at the photograph and he feels nothing from it, just the rings of passion that surround it.

The photograph returns. A badly glued edge clips against the base of the door, bends, threatens to snap, survives. Howard is glad that his son realises how important the photograph is. He is sure his son is nodding. His son will be waiting. For what? Something more than explanation or apology.

— I want you to be proud of me.

Did he say it or just think it? Howard stretches his torso over each leg in turn. His back had been getting stiff. He puts the photograph away in his jacket and now will have to find something, perhaps sandwiches or a pen, to store in his case. He wants to tell Simon about adventures he had in Cornwall, about the affair he had with a barmaid, about the game they used to play — imagining journeys together embarking from the house he rented overlooking the harbour. It's a tale of love and blood. That's the kind of meat the boy will have a taste for.

Sentimentality and a slow train took Howard to Port Jacob. He rented a white weather-boarded cliffside house on the Atlantic Ocean. He listened to records a lot of the day. He went on blustery cliffside walks. He acted the part of the courteous stranger. He felt politely invisible.

He made one friend, a handsome, wise old fraud called Geoffrey. Geoffrey was the rheumatic lover of the absent owner of the cottage Howard was renting. A retired civil servant from London, he had successfully assumed the persona of a sea salt that charmed the tourists, and was also the backgammon king of Port Jacob. Howard sometimes beat him, usually lost, and he paid for his losses with pints of local ale. They played most afternoons on the empty pub terrace, Howard in Somerset farm clothes, Geoffrey in his blue skipper's cap, heavy blue fishing jumper, wild white beard, heavy golden band through his right ear-lobe. And in between games Howard ogled the barmaid.

The barmaid smoked roll-up cigarettes. She was sturdily, boyishly built, narrow hips, strong jaw, shoulder-length hair dyed a deep dirty red, electric-blue eyes that made people think there were dangerous thoughts sparking through her head. She was the local siren and was not, Howard had failed to realise, touchable. When the pub was quiet, when the tourists were sitting quietly on the terrace overlooking the harbour, when the local old lags were playing angry games of dominoes, when the men of the lifeboat crew were singing sea shanties and Beach Boys songs in their enormous baritones, Rachel would smoke her roll-ups by the side of the bar and young men from the village and beyond would lean against it on the other side and offer her their hearts, and if they could make her laugh she would let them continue to try.

Howard was bored in Cornwall. He decided to make friends

with the red-haired barmaid. But, unwittingly, he committed several punishable crimes. The first rule he broke was not to stand in the appointed courtier's place by the side of the bar. He sat on a stool by the middle of the bar, face-on to Rachel, his head ducked beneath fishing tackle hanging in a net from the ceiling.

Howard's second crime was not to make a lurid, lavish, humorously base appeal for a fuck that could be acceptably dismissed with a look and a curse and a giggle. And these crimes were not going unnoticed — a young man, the brewery manager's son, whose pale face was decorated with a weak moustache and a lazy left eye, was impatiently waiting in the suitor's appointed place by the side of the bar.

Howard had not tried to make the barmaid laugh and he had not appealed for a fuck. He had smoked cigarettes and drunk pints of beer and told the barmaid about the trip that he was planning to New York. And then, with no flourish of lasciviousness or gallantry, he had left the pub (nods and smiles to the people he recognised along the way — the postman, the butcher, the woman who worked in the general store, the man who ran the post office, the drug dealer, the lifeboat crew, the brewery manager's son) and returned to his cliffside house where he watched the ocean tide sparkle and the lights of the village twinkle and die.

Howard was bored. His hair was a joy to him, crown fully covered now, with no grey to spoil the perfect blackness. But Howard was itching for adventure and, like a mad scientist with his deformed slave or a stern-jawed hero with his soft-limbed sweetheart, he needed the help of another to build the temporary door for his possible selves to saunter through. He was full-haired, almost friendless, and bored. He went to the pub every evening before and after dinner, which he ate by turns at one of the four harbourside restaurants and on the fifth night at home. Rachel worked a mysterious set of shifts he never did learn to predict but she was there most evenings, and the next time Howard went to the pub he broke the third rule: he invited her to come back with him to his house at the end of the shift.

She refused, partly out of loyalty to the local laws of courtly romance, partly because she was frightened of him, partly because she wanted to accept. Howard took it in good heart with a monkish gesture of infinite acceptance (a clasp of the hands that looked also like the beginning of a golf swing). He left the pub, walked down to the beach, where he sat on the long wooden racing gig, keeping cold company with a family of bickering seagulls and an enthusiastic little black dog that kept yapping inefficiently after the unconcerned birds.

The night that Rachel consented to come back with Howard to his cliffside house was her birthday. There had been songs in the pub, and a birthday cake topped by two dozen trick candles that relighted when blown out, and Rachel had been hugged by each of her courtiers (and twice by the brewery manager's son), and kissed by all three of her best girlfriends. The pub stayed open long after closing time. The lifeboat crew sang its songs. A middle-aged blonde was helped up on the bar where she performed a giddy, skirt-lifting dance. Blushing Rachel smoked her roll-up cigarettes. Her suitors made modest stabs at her with their open hands. And when the night was over, and the street outside was full of pleasantly drunken mobs of neighbours and friends reluctant to go their separate ways, Howard was walking towards the hill that led up to his house; it was dark, no moon, and Rachel was walking bashfully beside him.

They spent the night in the living room, ancient pop song soundtrack, Syd Barrett, Eric Burdon, bearded Beatles, the Atlantic wind knocking at the loose windows, as they kissed each other briskly, almost frantically. Sometimes one of them would doze, sometimes the other, and she asked him to hold her very tight in a whisper borrowed from the movies.

The next night Howard spent an unnecessarily long time talking to the brewery manager's son. His name was Dennis and he had wanted to be a maths teacher but family pressure had heaved him out of university into the service of the brewery. Howard rather liked Dennis, he liked the resolute way he always wore the same two badges (I ♥ BEER and PORT JACOB ISN'T THE END OF THE WORLD . . . BUT YOU CAN SEE IT FROM THERE) on the lapels of his denim jacket, but Dennis kept talking about his lost mathematical career in a way that felt significant (made more so by his wandering left eye) but never conveyed its full meaning. Until the final night in the village Howard and the brewery manager's son did not speak again.

There was talk in the town. The hairs on the back of Howard's vulnerable neck bristled with the sly anger of other people's eyes. The woman who owned the general store and ran the charity events was particularly cold towards him. It was only on his last, violent night in town that Howard was told by Geoffrey that this was Rachel's mother — he should have guessed it, the same clumsy application of hair dye, the similarity in figures; the difference between them was that Rachel was accidentally beautiful. Only one person ever referred to the affair and that was Geoffrey. They were playing afternoon backgammon on the otherwise empty pub terrace. Geoffrey was winning and Howard was mildly joking about it and Geoffrey said something about playing dangerous games with children which Howard asked him to repeat but he would not and the game ended early and bad-temperedly.

Rachel was scared to leave Port Jacob. Once, several years earlier, she had gone away to art school in Plymouth. She had lasted two miserable weeks there, returned home, and never dared to escape again. She loved to talk to Howard about travelling. Her favourite game, and it happened to be one of Howard's favourite games too, was to plan journeys. The two of them, in the smallest hours of the morning, would sit on the living room floor in Howard's rented house, with an atlas between them. One of them — usually Rachel — would open the book at a random page. Then they would invent their routes, pronouncing the names of the places they would stop at in holy hushed tones.

The only time they argued was when Howard would not shift away from America (pages 36 to 37). Rachel wanted to return to India (58–59) and particularly the south-west (in detail, page 62) where she had enjoyed herself on sunny opium beaches two nights before. Howard was investigating the Atlantic coast, from Virginia down to Florida. He was imagining himself on a fishing schooner chasing marlins and sharks, Captain Howard, the gorgeous pirate, glamorous and unrequired.

Rachel grabbed at the book, Howard pulled it back. Rachel caught an edge of page 37, Howard yanked it away: the page tore, New England to Labrador ripped out, briefly free; it flapped and then swooped to its death on the carpeted floor. Rachel looked at Howard. She was anticipating something harsh, she was prepared to cry. Howard smiled, Rachel smiled too.

One evening Howard tried to persuade Rachel to leave.

— Just go, he said, you've got money saved. I'll lend you more if you need it. Don't worry for your parents or for me (if you were going to). Take a train. Break out of here. Come back the same day, the next afternoon, another year. Go to a place with a name that excites you. Disco, Tennessee. Normal, Illinois. Toad Suck, Arkansas. Macao. Goa. Do it now. Soon it will be too late.

— I would like to, she said. (Shy blushing girl, pretty strong-jawed face lowered, eyelashes flicking off little damp spots of mascara, her gift to the floor.)

— Then do it.

— I would like to. I can't.

He made her give reasons and she laid them out in a line and he demolished them coldly one by one. Her mother, money, her father, the future, her sick dog, the surfing beach, her fear.

He lit the cigarette she had just finished making. He turned over on his back and stretched his arms and legs out as if he was making an angel in the snow.

— Take the atlas. In a year's time I'll meet you in New York City. That's where I want to be.

— Promise?

— Promise. On your twenty-fifth birthday. We'll meet in New York. At, at the top of the World Trade Center at noon.

It was where he was meant to be, New York. Where Grandfather Jack Levy thought he had disembarked but instead it was Tilbury, home to evil, dock-lounging Lenny Leverton. Their alternative family history would begin in New York.

Rachel left the next day. Howard watched her from the balcony of his cliffside house. She sneaked out of town with a suitcase carried in both her hands, with a rucksack misshapen by an atlas on her strong back. She walked away, up the hill, through the common land, onto the main road, taking stuttering steps, knees kicking the suitcase. She stopped once, just before she got to the bus stop on the main road, to rest the suitcase on the ground and rub her chafed hands and look, perhaps fearfully, back.

They wouldn't serve him in the general store that day. He waited by the counter with a bottle of cider and the woman working there, the owner's friend, a bitter saintly virgin with a prominent white moustache, served everyone else, managed never to look at him. And when the shop was empty, and Howard was still there, holding his bottle of hand-warmed cider, she turned her gargoyle face to his.

— Do you know what you've done?

— I —

— Do you *know* what you've done? She's run away. Everyone's driving around looking for her frantic with worry.

— You should give her a chance.

— Fuck off.

The owner's friend hissed at him like a snake getting ready to strike. Her moustache trembled. Howard put down the bottle of cider, returned to his house.

They were waiting for him later outside the pub. The brewery manager's son and four strong boys with him, two either side of the door, sinister night ruffians in jeans and T-shirts with bland expressions on their potato faces.

— A word, said Dennis, the brewery manager's son. Let's walk.

They walked. Dennis and Howard side by side, two night loungers in front of them and two behind. They walked up the lane to the general store and went round the side to the promenade that overlooked the beach. Dennis leaned against the promenade wall, his back to the ocean. He patted the top of the wall beside him. Howard took his position there. Their shoulders touched. Both of them looked straight ahead, at the white shadowed side of the general store, at the four strong boys nonchalantly lighting cigarettes and throwing their matches against the wind towards the ocean.

— You're a sneaky fucker, said Dennis, sadly.

Howard opened his mouth to speak. He wanted to assure

Dennis that there was nothing to be jealous about, they had only planned journeys together, really that was all.

— Shut up. You'll get your chance. Let me speak and then you can speak and then we'll see where we stand.

Howard shrugged. Dennis's shoulders moved up and down with his.

— Now look. Cards on the table. Things to say, let's make it plain. So that you understand. No. Don't say anything. Let me speak. I'm going to tell you about the trees, something I've never said to nobody before in my life and you're going to hear me out. And then we'll see where we are. We don't have a lot of trees around here — and they're small trees at the best of times, you probably despise them, but that's how it is because of the ocean, the salt in it, and the winds, they do their best. This is the fact now, it might embarrass some people to tell it, not me. I've carved our names, or at the very least our initials at least, into every single fucking one. The first I did when I was five and that was twenty years ago. I'm not saying this to make you feel sorry for me. There's one tree, it's out at the school field on the top, I kissed her by that tree once (had to give her fifty pee to do it but that's not the point), and to me, you understand, that's always been *our* tree. I go visit it sometimes when things are bad, against me — three of these boys work for the brewery by the way, don't worry about keeping them up, they never get bored or cold; and Tony's a copper so he'll put up with anything — When I came back here after I stopped with the maths what do you think the first thing I did was? I went to that tree. Just stood there. Magical thing, the tree was bigger and our letters were bigger and the heart I carved was bigger too and that made it fucking magical. Do you despise me?

— No, said Howard. I do not despise you.

— I don't care if you're lying. If I'd become a maths teacher I might have been able to take her away to the town, Bodmin, Wadebridge, maybe Exeter even. I know she wanted to get out of here, I knew that. I'm not stupid. I know. But it was too difficult for her. She's scared of things.

— Maybe she just believed it when people told her she should be.

— Is that what you think? Really? Well get this. I don't give a fuck what you think.

For a moment he was silent. Howard could feel him either nodding or shivering. The night was getting cold. The four strong boys were busying themselves with a stone-throwing contest. Now they were squabbling over whether run-ups were allowed or not. Howard liked Dennis. The strength of his emotion was admirable. Enviable.

— Where is she?

— I don't know.

— Where is she?

— I don't actually know.

— You don't actually know. You don't actually know, actually. Fuck off. Tell me. Actually.

— She's gone away, that's all I know.

— Come on. Don't play me for a fool. You're meeting her somewhere. Where? Plymouth? Bodmin? Truro? *London*?

— I'm sorry. I appreciate everything you're saying. I do. It's very . . . loving. But I honestly don't know where she is. She's gone away.

— Look. If you're trying to avoid getting beaten up don't bother. Save your fucking London toffee-nosed breath. You're going to get the shit kicked out of you. That's what I call a given. Me and my friends here are going to punish you. We're going to punish you because you came into our town and you're a wanker and didn't have the heart to look after her properly. And no one's going to come this way and even if they did they wouldn't do anything because we're in the fucking right on this and everyone hates you here anyway. You're meeting her somewhere. Where are you meeting her?

Howard didn't say anything. He couldn't think of anything to say.

Dennis stepped away from the wall. The strong boys threw their final stones and flexed their strong naked arms. Howard closed his eyes and lowered his head and felt Dennis's gentle fingers cupping his chin.

— You fucker, said tender Dennis.

They hit him and they kept hitting him, in the belly, on the arms, the face. Howard found it was easier if he kept his eyes open. The worst thing was the surprise. It helped somehow if he knew who was hitting him and where. They took it in turns, the polite strong boys, with open hands and closed fists and booted feet kicking him in the shins and the groin. Dennis tried to hold him up but Howard was too heavy for him and rolled slowly down the wall, lay on the cold ground, curled, his back to the wall, his knees protecting his groin, his arms hugging his head. It was very intimate, the attentions of these boys on his body. He admired their single-minded strength. And when it was over and he was lying on the ground, looking up to the stars under bruised eyelids, he heard the boys' loose panting breaths and the scraping of matches to light cigarettes and their footsteps slipping away.

And Dennis was still with him, sitting beside him, a managerial hand resting on Howard's damaged chest.

— Where is she?

Howard tried to shake his head, he tried to answer but that was too painful too. He found that he was able to smile, perhaps for the first time in his adult life, and he did that, held a pained, cracked lip smile open to reveal his broken front teeth.

He bathed his wounds at his cliffside house. He covered up the cuts as best he could with sticking plaster and rough patches of gauze. He was sure his ribs were broken. Whenever the pain made him feel like crying he would stop dressing his wounds and just rest his palms on top of his unblemished hair.

He wrote a letter that night. Howard Levy again, he wrote a sentimental letter to Vivien about the past and the future and money and the children. And even though he tore it up and scattered all the pieces to the wind, he paid for the letter later that night, with a clump of short dark hair on his blood-spotted pillow.

Howard sits on the cushion and his arched back feels like it's breaking and he tells Simon the story of Rachel and the atlas and Dennis the brewery manager's son. Simon likes the story. Howard knows he likes the story because he can feel what happens to the emotions contained inside it. They pour through the locked door and find their way inside his son and remain there, fermenting, intoxicating.

— I looked for the tree later, couldn't find it. The one he told me about, on my way out of town. I had the dental work done in Bodmin and then left Cornwall for good.

Does it shock his son? To hear his father tell stories of romances with barmaids? Simon does not seem shocked. Simon is silent. Holding his breath. Waiting for more.

— What did I learn there? That I wanted to be part of something I suppose. I have to go soon. I'll be late for work. Do you remember that patch in the garden? Where I caught you, when you were what — eight? You were off school because of some viral infection and you'd sneaked out of your room, bare feet, in your dressing gown, and I caught you spitting a load of phlegm into the flower bed because, you said, you wanted to grow a virus tree. Do you remember? The party was good last night, you should have been — No. It wasn't good. It wasn't. It started off well. I was rude to Tessa, that felt good. Rude just by telling the truth to her. I want to tell the truth. I want — does this sound silly? — I want to do, as they say, the right thing. Things.

He half-expects applause. There is no applause.

— What's that? I thought — No. Is that the time? I'll be late. Simon. That's enough for now.

His voice sounds like Dennis's, aggressive and uncertain, which makes it easier to say what he needs to say.

— I'm going to make you fucking proud of me.

He thinks he can hear a sigh but he can't be sure. Howard stands stiffly up, picks up the tray of his son's breakfast things and carries it down to the kitchen.

Vivien is there. Of course she is, washing up with yellow rubber gloves, an apron on to protect the navy skirt she wears and the white T-shirt filched from Howard's past. She cleans the coffee mug by touch and habit while she stares out at the garden. She does not hear him approach. Is she back there now? In the world her letters made? Ruthie will be choosing a medical school, Simon surely must be about to go to Oxford, and who will be providing now for Vivien's needs? She had disposed of fantasy-architect John, had not convinced herself of the desirability of Mr Marks, so there was space in her world for a new sex-glamour god.

Howard puts down Simon's breakfast things on the work surface beside the sink. Vivien is startled. She takes a step back, a rubber-gloved hand goes up to rub soap bubbles against her suddenly blushing cheek, she blinks at Howard several times as if she cannot be quite certain what he is.

— Howard. It's you.

He nods. She nods. They nod at the same time. On this point they can both agree. She waits for him to touch her, for him to find a part of her body in an action that she will endure or rebuff or perhaps enjoy.

He wraps his arms around her shoulders. They stand chest to chest, apron to shirt. She holds her arms outstretched to keep them from getting him wet. There are still times when hers is the most excitingly familiar warmth. He kisses her on the forehead, the cheek, the mouth.

— I left my case upstairs.

— You had better get it then. You'll be late. I've made you a tuna fish sandwich.

That's true as well. He leaves her standing as she had been when he arrived. Washing up, gazing upon the garden, as if the flowers are her imagination's secret trampoline.

He takes the stairs two at a time, towards the sound of a voice spilling secrets.

— . . . Of course Andy was awful. Sleazoid fuck, I knew that at the time. He's probably gotten down to the fourteen-year-olds now. He was horrible, a chicken fancier lech with this old man skin that, like, sags. And he'd go on, just go on, about how things were in his glamour days, nightclubs and bands and stuff before it all got tacky. It was so boring. You should have heard him. And they're all like that, they've all got some *thing* that they can't stop talking about. Everyone. Animals or bands or diseases or computers or bikes or something. They've all got something to be boring about and I don't even have something that I can be boring about . . . No one ever liked Andy. Me neither really. Blond has never done it for me, you know that.

She sits, his daughter does, in one of Howard's old T-shirts pulled taut over her upright knees. (And Simon too? Does he wear one of Howard's abandoned T-shirts? This their domestic uniform?) One hand rests upon Howard's attaché case. The other picks flecks of mascara away from her eyes. She pulls away from the keyhole when she sees Howard come near.

— Dad. You look nice. I want to come with. I've got your case here, I found it.

She offers it to him, he takes it, tries to ignore the look on her face. Beseeching.

— Not today, honey. Too much on. It's my first day back.

— I used to come. You used to let me.

She did, in that period that for her must feel charmed in retrospect, when she could walk and Simon couldn't and she would act so tough and in charge of things and people, particularly her father. He used to take her to the factory on school half-term holidays. Put her in the same office he'd been put in when he was her age, give her paper and crayons to kill the time. And when he'd come down to look for her she would always have gone wandering, charmed by the smell of workshop grease to the shop floor.

— You were young then. Shouldn't you be doing something? Looking for a job?

— Daddy!

About to turn. If he pushed it a moment longer she would become petulant and evil. If he tickled her or reminded her of something childish then she would giggle and be a child again. She squirms on the carpet. He tries not to look at the things the T-shirt does.

— I have to go. We'll talk later. I'm sorry.

— Yeah right.

He takes the attaché case into the bathroom with him. He rests the case on the bidet as he pisses. Picks it up again as he inspects his face in the medicine cabinet mirror. When he goes back out and down the stairs Ruthie is deep in confession again.

— You always were foul. I don't wonder you hide yourself away. Look. Listen to me. Are you by the door? I might even tell you some of what happened at the party, and I might even tell you some of the things that I *felt* ... And — come closer, turn down the radio, I'm going to have to whisper —

I've got something nasty to tell you about Uncle Leslie. That's better . . .

SIX

OLD MACHINERY

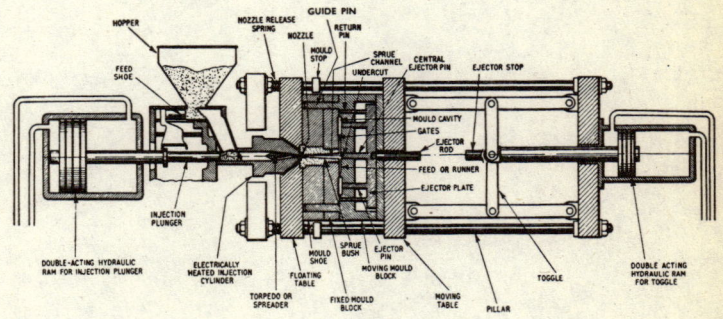

THIS IS CHARLIE'S world, not his. It probably never was. And it is sickening to enter it, like a playground filled with unfamiliar wars on the first day at school. He wants to run back home, call out to Ruthie, *Yes you can come with me, dress for sex or whatever it is you put your clothes on for. And bring all your friends, whoever, I need all the help I can get* . . . Instead it is just Howard alone, parking his car in an unmarked space, taking as long over locking the door as he can, walking through the car park to the workers' entrance (he's always gone in by the workers' entrance, that's how he's shown his solidarity over the years, the reception entrance is for customers and bosses, not for men who know how to get their hands dirty). He wants to run away, he wants to cry. He wants his mother to surround him with breasts and arms, but she is dying, yellow, wordless, in a TV-illuminated hospital bed. He walks on, employees he recognises show no sign of recognising him. The gun-metal door, is that Tony walking ahead of him, who he used to talk about the golden age of reggae with? Probably. And if it is, that's Tony who fails to hold the door open for him — so Howard has to jump suddenly forward to catch it (a thought whirling through his brain, a child's compact with fate, that if he allows the door to close nothing here will turn out right), he stops the door in time, a few hairs flick down past his eyes and are gone; he feels the need to be sick now, this is awful, intolerable, unendurable; he enters.

Later he will try to tell Simon about it. He will try to tell Simon but he will only find silence because the words he needs to use are not there. The smell of the corridor leading along to the shop floor. (He will say sweat and oil but they won't be enough.) The taste of spring water from the drinking fountain in the corridor. (Metallic? Acrid? He thinks it tastes of childhood but he'll keep that to himself.) He stays bent over the fountain, workers going by ignoring him as if he were an unthreatening ghost of the past.

There's Charlie. He can see him through the semi-frosted glass of the reception area window. Howard waits for Charlie to leave. Charlie doesn't leave. He's performing one of his routines, a shtick that seems to involve two women, another man, a cigar, a mug, himself. Charlie looks bigger than he used to. So does everything else. Howard the mouse patters tinily along. He can't face the offices yet, the secretaries, Leslie, Charlie, Charlie above all, so he takes the stairs down to the basement.

Howard remembers when this had been The New Building so how can it now be so dustily lit and smell so bad of age and piss and decay? He'd come down here on the day it had been finished, he'd walked along here alone, a fatherless inheritor in short trousers testing his fingertips against the paint-tacky walls of his future kingdom. Howard kicks past store rooms and junk rooms and an employees' room where broken moulds and damp cardboard boxes containing the unsold toys of a generation ago lie against the spidery cracks of porcelain urinals. An inspiration hits him. A memory so clear that he worries he might have dreamed it or stolen it from somebody else. Howard stretches both arms wide out and he walks slowly, closed-eyed, along the corridor with his fingertips gliding against both walls: no joy, not here, just pocked walls and light switches and splintery door frames — and when his forehead slams against the bolted fire escape door he realises he's been looking too high and painfully he turns around: he walks back to the stairs crouching, his knees ache to do it, arms still out, an arthritic Cossack dancer with an albatross wingspan, and finally, beside a toilet door, beneath a round scar in the wall that feels like it has been rammed with something round and metal, he finds what he has been looking for.

Howard places his adult hand against the child's print on the wall, the secret mark he had made back then, when this had been a quiet new green place that smelled of paint and turpentine and the future. He had pushed his palm hard flat, whispered

262

a quick vengeful prayer, dug in with his fingers to make his lasting impression, then he ran back upstairs to rejoin the party and surreptitiously daub its edges green. This was the mark of his hand when he was eleven. It looks stronger than he remembers it. A hopeful message for one of his future selves. Howard returns upstairs.

Charlie has gone. Just the two secretaries (is he still permitted to call them that?) half-standing out of their orthopaedic chairs. Everything here is so confidently substantial without him. Even the office furniture exudes gravitas. Barbara moves away from her desk. She has put on weight, hard to imagine the two of them had, at a Christmas party, almost. She offers him a drink with a dainty arm movement to accompany it and a shy smile following. *I've saved your mug for you, here's your coffee, how you like it, milky, one sugar.* Howard doesn't have the heart to tell her that he takes his coffee differently nowadays, sugarless and black. He accepts the mug (HE'S THE CHIEF! grey cloudy letters like campfire smoke, cartoon red Indians bowing cartoon obeisance), hits her with his new smile, and then the other one — *Valerie!! How are you?!!!* — little simper from secretary number two (or is it personal assistant? maybe office manager? Could even be director of administration, have to ask), he pushes on, through the swinging wooden partition, hesitates as he reaches the stairs. Then he walks on, head down, eyes counting the steps; if he reaches his office without having to talk to anybody else then maybe this might turn out all right.

He has returned in freedom. He tells this to himself over and over with his office door closed and his desk drawers open. He was made to run a gauntlet, of shakes of the hand and *welcome back*!s and little cheek-kisses until, reaching his office door, he found himself alone with an unlabelled bottle of champagne in his hands. That was half an hour ago. No one has sought him out since. If he chose he could go now, leave his office as he found it, disappear again into the miraculously functioning world. But Howard has resolved, Howard has chosen, in a Greek island harbour town, on a rickety aircraft, in a tube train, in his marriage bed, outside his son's door, to do the right things.

Prominently on the desk, in front of a computer terminal and keyboard, he has placed the photograph of Daniel and Samuel; the metallised polystyrene frame they are mounted in is a poor reward for the damage they have endured. Next to the picture is the telephone. After a couple of tries Howard works out how to buzz down to reception.

— Barbara?

— Valerie.

— Valerie. Sorry. It's Howard. Could you ask Charlie to come in to see me? Thank you. Sorry.

He could of course get up himself, open the side door, walk past their shared washroom, knock on a second door, push it ajar and invite Charlie in. But it seems more professional this way. Howard, in control, waits for Charlie. He waits for the battle to begin again. Howard clasps his hands together on the desk, a statesman's pose. He clears his throat. He squints to read the inscription beneath the display button mounted over the mahogany mantelpiece. Even his eyes are failing him now. A noise in the corridor outside. Howard clears his throat again. He looks down at the desk and then up at the ceiling. Someone walks slowly past his door without stopping. He has already been forgotten.

It would be foolish to pretend to be a statesman for an audience of none and then die locked in the pose, to be discovered later by an unknown descendant, one of Ruthie's children by any number of muscular unsmiling men or maybe a raffish son of Simon. *Who did that use to be?* they would ask, pointing through the dust to the dead man at the desk, his flesh hanging down in ribbons over the decayed cotton of his pinstripe suit, his ivory knuckles still clasped statesmanlike together, his skeletal mouth held in a firm, not unattractive skull smile. *Oh that*, the answer would come, *that was your grandfather, he went missing one day, closed his office door behind him and was never seen again . . .* And then a child's finger would come boldly out, *May I touch . . .?* and before the reprimand could come the finger would penetrate cotton and flesh and bone and the whole corpse would crumble into dust, the powdery sound of it lost beneath the sobbing of a suddenly terrified and remorseful child.

264

Howard gets up. Howard goes to the corridor door. He opens it a crack. There's a boy in a grey suit running out of one office into another. There's an unfamiliar maintenance man standing beneath a flickering neon strip light with his arms akimbo and a chiding expression on his face. There's nice, trustworthy Mr Katz, can see the back of him retreating down the corridor. And there's one of the marketing people (Gary? Steve?) advancing past Katz, past closed office doors, and past Howard without even a glance.

Howard shuts the door. Howard returns to his desk. Howard will assert himself now. Howard pushes his weight back against the chair so it wheels to the wall. Now he will be busy. Once he squeezed elements out of his life. Now he will decide which to squeeze back in. Howard sifts through the papers in the drawers: internal memos, consultants' reports, stationery, paper clips, computer manuals, a guarantee from the telephone company, other people's work. Howard will rebuild his corporate life piece by piece. Choose each item that is allowed inside. That report, a feasibility study on a new customised computer system, does he need it? Hardly. The system was installed years ago, nearly bankrupted the company. Throw it away. This business card? That airline timetable? This photograph of a competitor's Birmingham plant? This letter from a prospective buyer of the firm? That toy brochure? This poem written in longhand to fill a school notebook? *Poem*? The telephone buzzes.

— Sorry Howard it's Valerie.

— Hi Valerie.

— Hi. Charlie's gone for the day. He left a message for you on the computer.

— Great. Thank you.

Fuck Charlie. Foolish Charlie is underestimating him. He's allowing Howard a clear run at this. Howard will find his place here and then Charlie will be sorry.

The poem, tiny italic scrawl inside a school-blue rough book. This is Joe the Poet's great work, *A Justification*. It might even be the only copy, Howard can't be sure. Excerpts from the poem were published every half year in the same dreary magazine. There had been a moment though when the half-year had come around and Joe had not composed enough new lines to submit. Howard can't remember what happened after that. Either the magazine folded or Joe did. Both came to pass but he can't remember the order they passed in. He throws the book away.

In the drawer below, Howard finds a prototype sketch. No signature but it is unmistakably designed by Danny. Here are the familiar tail loops of his father's *f*s and *y*s in the lettered annotations. Howard tries and fails to capture his father's imagination through the foreign lines of the technical drawing. Instead he reads the process description on a companion sheet. It must have been a frivolous moment when Danny designed this. Here is a machine to produce self-decorating candles. As the candle burns, different coloured waxes run down prettily planned routes. Perhaps that's what the world needs now from the Levy family. A self-decorating candle built with ancient technology. Howard hesitates. Howard throws the plan away.

Brochures. He throws them away. Questionnaires. He throws them away. (Howard had been proud of his questionnaires, the Howard Levy Personnel Psychological Profile.) Management reports. The drawers are silly with management reports. Before Howard had run away he had gone a little mad hiring consultants so he could then ignore what they told him to do. There was one Trotskyist who advocated permanent revolution in staffing policy. Another was a games theoretician, another big on chaos theory and Gaia theory, mixed and matched to provide a set of rules for the perfect running of the perfect modern firm. The only thing that all the experts had agreed on was that there was something rotten about the company. It needed rescuing. Howard throws the reports away.

Howard throws away everything that is superfluous. Howard throws away hair oil and keratolytic pills and an appointments diary. When his desk is empty and before Howard attempts the computer, he fishes out Danny's sketch and Joe's poem from the bin. He looks quickly around as he wipes them down, and then he places them carefully away in the bottom right-hand drawer. He shuts the drawer quickly. He sits up at his desk like a statesman. He smiles for an imaginary camera. He hopes his weakness for the past has not been observed.

2

HOWARD WAITS FOR his turn at Simon's door. He wants to tell his son about his first day at work, the confusion of it all, the wandering around the executive offices, the picking up of things and the putting down of them again. He had spent a long time at the mirror in the washroom he shares with Charlie. With a pocket mirror angled behind he examined the state of his head. Time travel. This was the head that belonged to Howard in his late twenties. The first indication of the hair hole at the crown. A decaying young man's doom.

He wants to tell his son too about what the day was like. He has told Vivien about it, but maybe because he feared she was bored and maybe because her imagination is a mystery to him he stumbled over the words. All the comparisons he found were empty, fanciful, foolish. He will stumble over the words again later, when he finally gets his chance, because he will be tired and the day has been long and because anyway he has not yet developed a gift for confession.

Howard waits in the kitchen. His wife is up there now. Several times he has gone up to see if she is finished and each time she retreated into silence as soon as she heard him near. And then when he bobbed up to her level he found himself fixed with a reproachful look — *I said I'll call you when I'm finished. I'll call you.* Then she waited, like the model of a wife in a Chinese family group the company had manufactured and unsuccessfully marketed years ago: kneeling, hands impatiently resting in her lap, gaze turned down, until he had had the decency to leave. She was beyond tricking. He had tried, stockinged feet on carpeted stairs, heavily away and softly back with stifled breathing, surely she can't hear him *now*... But each time she waited until boredom and aching lungs had driven him downstairs again. The pleasant, indecipherable song of her voice can only be heard from a floor away, never closer.

So Howard waits. He is sure his wife is talking about him. He wishes she had the strength or the need to say what she has to say to him, direct. Cut out the silent middle man, or middle boy. Ruthie is beside him, waiting for something.

Magnanimously, she has given up her place in the queue to Howard and now she waits for him to do the required thing in return. But she won't tell him what it is, just waits, eyes slightly narrowed, alert for the mechanisms of satisfaction or disappointment.

— What do you want from me, Ruthie? he asks, in his softest voice, but she shakes her head, how can she *tell* him? he has to *know*.

The wheels of disappointment turn inside his daughter. His wife still does not call down. His son is in his room. He has alienated his friends. It is hard not to accept the conclusion that he is alone.

— Howard!

His wife's voice, coming from the floor above. Howard gets up, kisses his daughter on the forehead, shakes some of the dead feeling out of his left leg (it had wrapped itself around the right and now he suffers for it), and hobbles upstairs. Vivien disappears with a book into the bedroom. She reappears to smile at him, a smile that promises she will be asleep or at least pretending to be by the time he comes to bed. Howard makes his way to the sitting place by Simon's door. He will talk, hesitantly, hoping that maybe in the process of confiding he will discover what it is he wants to tell.

— How are you? Stupid question. Have you finished with your tray?

Howard settles in. The cushion adjusts to his shape. The wall opposite is already familiar. They sit in silence for a while.

— My first day back. You'll want to know how it went. I survived. It was, really was awful to begin with, I was like a ghost in the corridor, the place smelled of sweat, oil. I didn't know what to do, how to act, how to start. I drank some water from the water fountain, the water tasted of, I don't know, metallic I suppose the word would be, acrid.

Howard can almost hear Simon asking him to go on. *Yes? And then? And then? Daddy? Talk. Talk.*

— I ate lunch by myself in the canteen. That was strange too. They made all this fuss when I arrived and then no one joined my table, not even Leslie who always used to plague me in the old days, sitting opposite trying to start one of his, you know, Interesting Conversations. The queue to get out of the car park at knocking-off time. That was the same. Or maybe it was a little less crowded. Or maybe more, I don't really remember. Charlie has a driver now. Someone who Ruthie used to know. Charlie. It's all about Charlie. Didn't talk to him but he hung over the whole day. He still has that stupid plastic car. Just pays someone else to drive it for him. And it's not even his real name, did you know that? Real name Lawrence, for your great-grandfather, Dinah's father, Fischelis the furrier. He was quite successful, had a house in Hampstead, knew Rothenstein the painter. Charlie dropped the Lawrence when he was a teenager. I'm Charlie now, he said. Thirteen years old, ugly as sin, bursting with hormones. Pimples and greed. They all thought it was because he didn't want to have a Jewish name. So why didn't he drop the Levy then? I'll tell you why he did it. To escape the curse of Lenny Leverton. Lawrence Levy. Lenny Leverton. Double *L*s. Our family's hate figure. Charlie didn't want to get cast in the role. The conman who will destroy our fortunes. He's up to something now. I know him well enough to know that.

Maybe Simon knows what it is. Maybe Charlie comes here, in the dead of night, to unburden himself of nasty shameful plans. Perhaps Simon is silently miming them now.

— I didn't see him. He left a message for me, internal computer mail. Welcome Back. Have Fun. Charlie. Kiss kiss and a smiley face.

— Tomorrow I'm going to talk to them all. Everyone mentioned in that Buller Report. I'm going to uncover the truth of it, find out how the whole fu — , awful thing has got so bad.

And Howard keeps talking, addictive. Night falls and spreads. The house is quiet. The son must be silently huge with his father's emotions. Howard talks until his throat is raw and his mind is sluggish, and finally he stops, remains there, on the cushion, sitting, and then it is morning again, his wife had no need to barricade her chastity with books and smiles; he wakes, stiff, tired, his head aches and he is hopeful.

HOWARD LEVY, COMPANY man (acting), corporate investi-
gator, has two tuna fish sandwiches wrapped in cling film next
to the photograph in his attaché case. The Buller Report is in
there as well. In a few days' time, if he finds it in the attic, his
table-tennis bat will join them. Soon, at this rate, he won't have
to hide the shamefully few contents of his case with his body
every time he opens the clasps in public. Howard Levy drives
his company car to the company site. He notices little details
along the way, and tries to make himself remember them in
order to tell them to Simon later. (There is a tyranny in confession as well as a liberation.) The wild-haired male cyclist in
a woman's floral dress peddling fast down Frognal. The magpies
circling the traffic island where Regent's Park gives out to Mary-
lebone Road. That notice advertising the Polish War Memorial
ahead could be an instruction to clean it. Why are those children
playing on the dinosaur slide outside the Happy Eater painted
up as clowns? The dreariness of the car park. Wasted land,
could build something there. The defeated way the workers
amble towards the workers' door. He would like to chip at the
optimism inside himself and personally hand a piece to everyone. Send them out into the world, a Levyite diaspora — *The
company is dying now but you, each of you, can do great
things, if you choose to live a moral life . . .* If his mother would
recover her health or at least her mind then the world would be
a fine place.

They come to see him, the Department Heads, the ones whose
cold corporate fingers touched Mrs Buller and made her die. In
they come, none in mourning, their hair dressed with waxy gels
and styling mousses rather than ashes, their skin softened by

conditioning creams, adorned by kohl and scent. Howard sits at his desk, which has been freshly dusted and polished for the occasion. Sometimes he sits back, one leg rocking across the other, folded fingers making a cradle for his chin, the visionary; other times he leans forward, elbows on desk, right index finger pointing, the general. Or else he sits there sternly erect, the judge. He is giving them the chance to prove their innocence.

He will flatter them with questions about themselves, how they see their work, how they see the company — *Draw for me if you please, with words, your vision of the future.* And when they are lulled he will make them reveal the truth. The truth about themselves and the company they work for. That at least is the plan.

— The future is plastic superconductors, says Arnie. Arnie is American. Arnie wears round wire spectacles, thinning sandy hair, a V-necked yellow pullover piped with red. Arnie wants to move from Distribution into Development. His voice is thin and strangled-sounding. His face carries the expression of a zealot.

— What are your present responsibilities? Howard asks as he looks at Arnie and thinks, incredulously, *Is this my ally?*

— Sales file orders with me. I send 'em out. And then also I'm like, the janitor, I'm not complaining about this but it was never in the contract. Things go bust I fix them. Why? Because no one else does. The urinal blocks I unblock it. Same with everything else. There's a glitch in the computer programming? Give Arnie a call. The dogsbody. Comes when you whistle. I like to like, tinker a little on the side, lab stuff? Superconductors, that's the future, only one thing holding us back.

— What's that?'

— Plastics don't conduct so good. But they will. It's just a matter of molecules. Guy who solves this problem he's down in the record books. Give me lab time, that's all I'm after. We'll both be covered with glory.

— Do you remember Dover Soul?

— Yeah. Little outfit. We sent them a few machines. They received them. No complaints.

— And do you remember a company van breaking down, in

272

Dover Street, N16, two years ago? Ah, sorry (Howard checks his copy of the Buller Report open on his desk): Dover *Road*.

— Road. Street. What can I say? A van breaks down, somebody fixes it. I don't keep a diary of such events. It's great to see you back, Mr Levy. We can maybe expect a few changes? We could do with them.

— Maybe so. Send in the next person if you please. I've enjoyed our chat. We'll talk again soon.

John from Direct Mail is the subject of much graffiti in the factory toilet. (So are Charlie and Leslie. Charlie, according to most of the drawings, is a drug-sniffing penis with a bouffant hair-do. Leslie has a long hooked nose that drips like a tap and avaricious pound-sign eyes. Howard was pained to discover on his early morning reconnoitre that there is no caricature of himself above any of the wash basins or on any cubicle wall.) John, a mannish man with a short back and sides and a soft growling voice, is always pictured as a sexy number in a tight dress wearing lavish earrings and a fruit-bowl hat. In person he looks like a rugby player. Perhaps, Howard considers, that's the provocation, or maybe the give-away. *Is this my ally?*

— I trained initially as an accountant, says John. (There's a trace of Scots in his voice, does he know Miles Rodgers?) I feel, to tell you the truth, a little bored by Direct Mail. The creativity it offers is limited. I would prefer to be in Accounts.

— How would you rate Accounts' present performance?

— It's not for me to say. I would rather keep my opinion to myself.

— Please. It shan't go any further.

— So be it. Dismal.

— How so?

— You'd have to find out from them.

— I'm asking you.

— You will have to ask them. I could run Accounts and I would run it honestly. That's all I have to say. I hate Direct Mail. It's so hectoring.

— Quite. Do you remember a company called Dover Soul? It was targeted by Direct Mail. Hectored, if you like.

— I can't say I do.

— Dover Soul used to be a client of ours. It was still receiving

273

direct mail campaigns when it was already well in lieu of account.

— That's a list problem, isn't it? We compose the copy for the mailshots that get sent to places on lists compiled by Research. I have often argued that the list should be updated more often than it is.

— Why don't you update your lists more often?

— We are, I am afraid, in the hands of Public Affairs on that one. It's not that I'm trying to pass the buck, you understand. It's just that that is how it is.

— Of course. I understand.

Which is not entirely true. The more people Howard sees the more complicated things get.

The lawyer Katz is a dark little man with a languid air and a grey chin. In his antique three-piece suit he looks as if he's been born to the wrong time. His hair is peppery and sparse. There's a dubious intensity about him that reminds Howard of Joe the Poet. Howard suspects Katz of writing short stories.

When Howard was on the run Katz was the go-between, the only one permitted to know his addresses because Howard could be certain that they would never be passed on. Katz is the keeper of secrets, perhaps because he is too lazy not to. Presumably Katz keeps Charlie's secrets as well. It would be a blasphemy to ask.

— I'm seeing everybody, says Howard.

— I know, says Katz. I hope it does somebody some good.

— Do you remember Mrs Buller?

— Of course. Tragedy. Poor woman.

Katz shakes his head. Katz waits silently for the next question. Howard has always wondered if Katz has a private income. Katz is the only employee of the company who has never shown either any sign of malicious ambition or fear of losing his job. Even with Freda in the old days Katz was the only one who would dare to stand up to her when she was in one of her cold rages. But Katz too is implicated in the death of Mrs Buller. Katz is in charge of Legal Affairs. Legal Affairs sent Mrs Buller a letter the day before her drugs overdose informing her that the full thrust of the law would be used against her.

— An awful misunderstanding, says Katz.

— I feel partly to blame.

— Blame is a luxury you can't afford. Your father makes machines and sells products. And what do his descendants produce? Blame and guilt and polluted ground and bad air. Forget about blame. It's a hobby for lawyers and philosophers and married people.

— What then, Howard asks, can be done?

— Learn to become a businessman or learn how to walk away without ever coming back. Learn to be shrewd and learn to make money. Think about the future. Sack people. (Sack me if you like. Make an example.) Launch new products. Devote yourself to the possibility that things will work out well. Win the table-tennis competition.

— I'll try. Thank you.

The next is Nikki. She works in Public Affairs which used to be called Customer Relations.

— I love Public Affairs, says Nikki. I do. To make no bones about it I've built it up from almost nothing. I derive *enormous* job satisfaction from it. But God knows we could do with some help. Another two more staff at the very very least.

— Do you –

— The problem is, I spend half my life trying to mollify customers because of other people's cock-ups. Excuse my language but it's true.

Nikki (*Is she my ally?*) has packed several layers of foundation cream on top of her tan. Her hair is a sort of grey-brown colour. She wears a charcoal-grey trouser-suit, a cream silk shirt, unbuttoned enough so Howard can count the freckles (fourteen) on the upper part of her breasts. Her features are sharp and pretty. There is a light band across her left ring finger where a wedding ring might usually go.

— You used to be Freda's private secretary.

— That's right. After the old lady retired Charlie, I should say the other Mr Levy, the other Mr Levy gave me a promotion.

— Did you enjoy working for Freda?

— She was a lovely old lady. But could be a little strict and old fashioned at times. Oh dear. I shouldn't have said that, should I?

— It doesn't matter. I want you to be honest with me.

— I will. It's the best policy isn't it?

— It can be.

— I'm sorry. This is making me nervous. Being checked up on like this.

— I'm not checking up on you. Don't see it like that.

— I'm sorry. I'm very nervous.

Nervousness is making Nikki do curious things. It is making her scratch at the inside of her left thigh and it is making her play with the middle button of her shirt which is, Howard realises, one of only two buttons that are actually holding the shirt together.

— Don't be nervous.

— I'll try.

— Good.

They exchange nervously brave smiles. Howard likes Nikki. He decides that she probably is ideally suited for her job.

— When did Customer Relations become Public Affairs?

— About two years ago? (She looks at him with a nervous quick smile followed by a wince as if this is the important question, the one that will determine all their futures. He nods. She relaxes.) We thought it sounded more up-to-date.

— Two years ago. Would that have been before or after the Buller Report?

Howard regrets the question. It makes Nikki more nervous. He fears for the shirt buttons.

— *After*, she whispers.

— Why?

— Mr Levy. *Charlie*. He thought it would get us off the hook. That's what he said. And another thing? (She looks around, she examines the window for listening ears, the walls for silent cameras.) He's up to something with an art club thing. Gary in Marketing will tell you about it if you're interested. It's a scam. But please don't tell Charlie, excuse me — Mr Levy. He might have it in for me if he knew.

— As far as you know, did anything else happen after the Buller Report? Apart from your department changing its name, that is.

— I don't think so. But (she brightens up again, perkiness becomes her) it improved our efficiency by thirty-seven per cent.

If we had more staff for Public Affairs I can't tell you how efficient we'd be. It'd be good for everybody.

— I'll bear it in mind.

— You can transfer people from Legal Affairs. They're all crap in that department. And in Research. And Accounts and Billing. Marketing is good. Although that goes without saying. I think he's waiting to see you actually. Should I send him in?

— Please do. And thank you very much.

— Thank *you* Mr Levy. I'm very grateful for your time.

And off she trots, briskly, inelegantly, sexily. In comes Gary of Marketing, who used to be Gary of After-Sales. Crispy gelled hair, doubled-breasted suit, wide shoulders, wide lapels. *This is not my ally.* Maybe though Howard should pick up a few style tips here, this is how to dress, how to give the workers and the customers confidence. What customers? Howard has hardly thought about them. That, he supposes, will be next to consider. The order book and the products brochures.

Howard should ask Gary now about his work in After-Sales. He should ask him about the letter that his department sent (while Accounts was persecuting Mrs Buller for her non-payment of Dover Soul's debt) congratulating Dover Soul on its wise investment in the future, and enclosing a Customer Satisfaction Questionnaire. And then, he supposes, Howard should interview himself. He should ask himself about the Customer Satisfaction Questionnaire that he devised so proudly and he should ask himself about the meetings he organised with department heads insisting that the appropriate questionnaire be completed after every transaction.

Howard is tired. Gary is cocky. Gary sits forward in the chair and blisters Howard with his energy. Gary removes his watch, lays it on Howard's desk and promises it will belong to Howard if Gary wastes a moment of Howard's time. This is a trick that Gary has learned from Charlie. Howard inspects the watch.

— It's a very pretty watch. Thank you.

Gary then holds out a page torn out of a magazine. Howard puts down the watch and takes the page. Full-page advertisement on cheap paper. Hideous, small-scale reproductions of great art, The Last Supper, the Sistine Ceiling, Michelangelo's

David, modelled in plastic, and below ludicrous words, strident with capital letters.

— People buy this? Is this our stuff?

Gary grins. Howard hopes Gary and Nikki are an item. They would look energetically handsome together.

— It's a rip-off machine and I'm here to tell you how it works. Direct Mail wrote the copy. Charlie built the mechanism though. A few thousand grannies send their postal orders in, oh that would look lovely on the mantelpiece next to the stuffed head of their war hero husband or whatever. Right? The money's banked by Classic Club. Fair enough. But then you know what happens? Classic Club goes into voluntary bankruptcy. Right? The subscribers get their money back except the money is a cheque on a new company account and this new company is called Filthy Hot Porn Productions, yes? It's a lovely instrument actually. People who buy this kind of shit are not the sort of people who are going to like banking a cheque from Filthy Hot Porn. And it's only a tenner or so anyway, these grannies aren't rich but a tenner isn't enough to risk the looks they'd get from bank clerks and neighbours. Maybe ten per cent of the money gets cashed. And Charlie keeps the rest. He runs it through Accounts so Leslie is involved as well. Have I wasted your time?

— Thank you. Here's your watch back. I'm very grateful. What — excuse me for being blunt but I don't want to waste *your* time — do you want? Wage rise? promotion?

— Both. I'll leave it with you. Thank you.

Off Gary goes, reclasping his watch onto his wrist as he leaves the office with his rolling tough-boy walk. Howard feels exhausted to watch him.

Just Leslie and Charlie to go. It's late. Lunch has been and gone. He's missed afternoon visiting hours. Howard dials the hospital number. The telephone rings and rings and rings. Finally someone gruff picks it up and tells Howard to hold and a few minutes later someone even gruffer transfers Howard to the Robert Falcon Scott Ward. This time he is granted music to listen to as he holds. A tinny snatch of something classical and then the Irish voice of the usual nurse.

— Mr Levy? Of course. No change.

Of course.

— You will call me if there's any news. Please. Good or bad.

She promises she will. He doesn't believe her. He lets the receiver down. Tears are threatening to come. The company is like his mother. Equally mad, equally decaying. Katz was right. If it was a dog they'd kill it, and afterwards they'd congratulate themselves on doing the right thing. What does that say about his mother? *Excuse me docter, here's a syringe full of deadly poison, you might find it useful in your treatment of that woman over there . . .??*

— Send in Charlie now will you. Thank you Barbara.

— Hello?

— Barbara?

— Hello?

— Valerie?

— Who is this? Hello? What's your bloody game?

— Guido. Hi. It's Howard. Wrong connection. Sorry.

— Who?

Second time lucky. He buzzes shop floor's extension and Barbara answers.

— Send in Charlie now please. Thank you Barbara.

— It's Wednesday I'm afraid. That's Charlie's road day. Leslie I know is waiting to see you.

Let him wait. Howard would prefer to talk to Simon. There will be traffic jams on the way back home. That should relax him.

4

HOWARD CARRIES UP Simon's dinner. Ravioli and salad and a bottle of Mexican beer. He knocks on the door as he has learned to do, three raps, a pause between the second and the third, then Howard goes into the bathroom and waits a few dull minutes, as he has learned to do. He resists the impulse to cheat. When he returns to the door the tray has been taken inside. Howard sits down. The cushion is already used to his shape.

— Everything is worse than I thought it would be. Everything is rotten. It's like my mother. No control at the top. Different decaying organs battling against each other while the head is powerless to do anything. The place is insane. And this art club thing. It's pathetic. I thought Charlie was a bit bigger than that. Do you think Leslie's involved? What should I do?

Howard stretches. The wall opposite could do with touching up. The whole house probably should be repainted. Simon is quiet. No sounds come through his door. The boy will be listening, intent, rapt. In Howard's mind he has a picture of his son. It is a sort of amalgam between three faces: his father in the photograph, the frantic nine-year-old who can't remember the punchlines to jokes, and Charles Manson in 1969 looking serene as he is led away in handcuffs.

— One bloke I hired once told me if the person you're interviewing looks up to the left he's probably telling the truth and if he looks to the right he might be lying. (Or was it the other way round? It had something to do with eidetic versus conceptual memory.) Everyone I spoke to today looked to the right but I figured they were telling the truth most of the time anyway . . . I could probably get rid of Charlie for ever. Should

I do it? It's been a long day. Everyone hates each other. If I'd seen him during the day I'd have begged him to let me go. Here, have the whole shebang. I'll pay *you* . . . What do you think? I need an ally.

Howard has been hogging the confessional cushion again. Ruthie, feeling cheated and neglected and bored, stomps up the stairs.

— Come on, she says. There are others, you know.

— Look, says Howard. I'm talking.

— You always *are*. There's others. We want our turn.

— I'm sorry. I'm nearly finished. I'll call you when I'm through.

He waits for Ruthie to go and when she has he discovers that there is nothing more to say.

— Katz told me to win the table-tennis tournament. That's probably the best advice I got all day. I'll speak to you later maybe.

It would be nice if men still wore hats. Howard goes downstairs, calls out to his women that he is through. He puts on his overcoat and walks out, bare-headed, vulnerable-haired. The road takes him winding up to the Heath. It is dark. Some schoolchildren play moonlit football with crumpled-up blazers for goal posts. A tracksuited man with a face like a wolf is running a jagged path from copse to copse. A dog howls. Another dog howls in reply. There seems to be a bonfire nearby, he just catches glimpses of the flames through the trees, but when he makes for it it seems to have moved in a different direction and it is still just as far away from him and he is overlooking the swimming ponds. Five hardy lesbians serenely breaststroke their way across the ladies' pond. The other ponds are deserted. There is someone else watching with him, over on the opposite hill. A slim figure, a young man, sitting on the back-rest of a bench with his feet on the seat. Looks like he's wearing a balaclava helmet. While Howard's looking at the watcher, the night gets darker and the figure disappears inside it.

When he returns from his walk, Vivien is in the talking place. Howard approaches silently. She stops talking. He withdraws silently. She starts again. How can she tell when he is or is not

281

there? This faculty infuriates him. He takes refuge with a bottle of Scotch in the trophy room. He accosts his wife later. He's a little drunk on whisky and worry.

— Why, he asks, do you tell him things that you can't tell me?

She tries to pull away, he won't leave her be. She goes to lock up the house and he tags along, the kitchen door, the downstairs windows, the french windows in the living room. Everywhere she goes he goes too, pulling at her arm, trying to bully her or nag her into telling him something she feels.

— You can say things to him. Say them to me.

He says this over and over. Up the stairs, outside the bathroom door, then downstairs again because she pretends she is thirsty when he knows that she just wants to hide their almost-fight from the children. (Even though Simon is silent and Ruthie has gone out, giving herself in sex probably to someone who has already half-forgotten her.) Vivien pours herself an unwanted glass of mineral water. Howard persists.

— Say it to me, he says. Whatever it is. What you tell him. What you think.

— No.

Should he hit her? Earn the right to her confidences that way?

— I'm terribly tired, he says.

She wraps her arm around his waist. (Is that an inadvertent tremor of distaste? or just a shiver from the draught through the Xpelair?) She leads him upstairs to their room. He drops his trousers and shirt on the floor and climbs half-clothed into the bed. He switches the light off. She turns it on again. She sits on the side of the bed and brushes her hair and says, with her back to him, that she doesn't think that people can ever change. This to him sounds like a challenge. He pulls her down. He kisses curls of hair away from her eyes, from her shoulders, from her breasts — the last are his hairs but he kisses them away all the same, equal subjects in the democracy of marriage. He does what he has been afraid to do since his return: he makes love to his wife as if she was one of the women he fucked while he was away. And when their love-making is over and he watches his wife slide from freedom into sleep, he feels triumphant. Perhaps this is what they can't tell each other. Perhaps with bodies they can do better than words.

5

HOWARD HAS THREE tuna fish sandwiches in his attaché case.
Vivien makes him a sandwich each morning, he puts it in his
case, they stand close together, husband and wife, they kiss air
and sometimes cheek, he goes off to work, usually, like today,
stopping at the hospital to see his mother on the way in.
Howard doesn't like tuna fish sandwiches. Maybe once he used
to but he doesn't any more. He wants to tell her — *You see?*
he would say, *You see how people change?* But he doesn't know
how to tell Vivien he doesn't want her sandwiches without
appearing ungrateful. Neither does he know what to do with
the sandwiches once he has them. Eating them is out of the
question. Neither can he throw them away. He can't risk
the chance of someone seeing him doing it, report his treachery
back to Vivien, break her heart a little more. His attaché case
smells quite strongly of tuna fish now. So does the photograph.
He puts it back on his desk, its daytime home. An odour of
tuna fish clings to it.
  — Leslie to see you.
  — Send him in.
Leslie comes sniffling into the office.
  — Dinah?
  — No change.
Howard sits back, his visionary position. In the old days
Leslie had been his reliable ally, always to be counted on in
the event of a vote. Since his return Howard has detected an
unaccountable change in Leslie. He will have to be careful with
him.
  — Sit down, please Leslie. Shall I get you something to drink?
Leslie sits suspiciously on the tip of his seat as if the chair

will swallow him up if he relaxes his vigilance. His pate is wispy, freckled. His little brown eyes blink angrily.

— No. Nothing. Thank you. Please. Let us go on with the interrogation.

— Interrogation? Hardly. It's good to see you. I'm trying to get a sense of things, that's all. I know things haven't been especially easy while I've been away. Now I'm back I just want to get a sense of things.

— You are back now? For how long? Are you in? Are you out? Nobody knows with you.

— Leslie. Have I done something to offend you? I wish you'd tell me.

His cousin blinks more furiously. His effort to calm himself down is obvious, and embarrassing. Howard averts his gaze, pretends to be reading something on his desk, but all he can find is the Buller Report, the words of which he knows off by heart. When he looks up again Leslie is smiling at him and trying to look suave. The effect is frightful.

Howard had meant to begin with Accounts' part in the Buller death. He doesn't have the heart to push Leslie back onto the defensive, when all he can learn is that Leslie is incompetent. Of course Leslie is incompetent, but so is everybody else.

— Leslie. I'm interested in the future. In all our futures. Draw for me, if you please, your vision of how things can be.

Leslie wipes his nose and then his glasses.

— Vision? I hope we can all be a credit to those who went before.

That's it. Leslie withdraws into his silence again. There is a limit to how tactful Howard can allow himself to be.

— Quite. Anything more? No? I have to tell you Leslie. I've heard something a little disturbing . . .

— Disturbing yes. A family problem?

— A work problem.

— To do with . . .?

— To do with something that *might* be connected with your department. To do with, specifically, a fake company that was set up to market some art.

— I know nothing about art.

— I use the term liberally. It was a scam. A money-making

plan. All I want to know is for who? Who did it make money for? You?

Leslie stares at him. Howard recognises the expression. It is one usually saved for slugs and spiders and occasionally snakes.

— Excuse me. I do not like your gist. You are calling me perhaps a thief?

— No Leslie. I am not calling you a thief. I am asking for information.

— Information which I cannot give. So sack me then. Call me all the names you like. If my mother was alive she'd be spinning in her grave.

Howard sighs. It gives him something to do. Leslie coughs.

— Howard? I have a thought.

Leslie seldom has thoughts. Whenever one does come to visit, Leslie wines and dines it, takes it out dancing, buys it a brooch to wear and mid-price tickets for West End shows.

— And what is your thought?

Leslie licks his lips. The suspense is unbearable.

— I tell you, obvious and not so obvious. This company, you and Charlie, problems. Yes? Maybe not so much now, but used to be, and, God forbid, who knows? will be so again, personality clash, philosophical problems, not easy. And sense I get, neither trusts the other.

— So?

— Yes. Exactly. In ancient Carthage when problems with succession — old king dead, childless, mourning period, two claimants, princes both, of course, neither nominated new king by old king — excuse me, you follow? — power not necessarily given to older prince. You know why?

— Because it might make the younger prince a little jealous?

— Exactly. Very good. Bad possibilities. Look at usual system. One prince, older one most usually, becomes new king, younger prince, as you say, a little jealous, problem. Maybe civil war. Maybe older prince who's now new king must murder younger prince, for the natural good. You see? So, I hear you ask, what did they do? I'll tell you what they did. *Neither* becomes new king. Instead, third person becomes king. But special kind of king. Like a king regent. The princes are in charge, king regent has casting vote in event of a disagreement. And let me tell you

286

one thing, Howard (Leslie coughs, stares at Howard in an attempt at utter sincerity that just about comes off), in the event of a disagreement, I will always be behind you.

— And what about the succession after that? In Carthage?

— Older prince's oldest son is king in waiting.

— And isn't his position a little precarious?

— Yes. Sometimes. Often I think king in waiting met with unfortunate accident.

Leslie looks a little mournful. It does not make him happy to contemplate the death of princes. Then he perks himself up again.

— But not relevant. This is my thought. Ancient Carthage equals LPC. You and Charlie young princes. And me, perhaps — I say this with all due modesty — the suitable third person.

— So?

— So, maybe, I was thinking. (Leslie blinks, pulls at his nose, twists at his moustache, he is about to share the substance of his thought with the world and the prospect makes him nervous, protective.) You might be interested to sell your shares in the company to, excuse me, I'm about to sneeze, forgive me, no, the danger is passed, to me.

— Sell my shares? To you?

— Exactly so. And Charlie to do the same. Simultaneously, in case of mutual suspicion or misunderstanding. With adjustment of course, legal papers, so both of you keep same share of profits. And limits on what I can do, king in name only. Regent really. The two of you are in charge and little Simon, I would hope, the king in waiting. This is not ancient Carthage. I do not think there is danger of any unfortunate accidents occurring to him. This is my thought. Because, I tell you, I worry. I worry the two of you don't get along so good. I worry about the consequences. I worry for the company.

Howard reports the conversation later that day. He acts out Leslie's part, perhaps overdoes it a little, the fractured sentences, the rhythms of speech belonging to someone who has learned English late in life (whereas Leslie was born in Palmers Green and remained there until Freda took over control of the company and, with it, a suitably superior flat in St John's Wood).

He makes Leslie into a comic conspirator, the Gunpowder Plot acted out by Punch or, better, Judy, and has to break off every now and again to control his laughter. On the other side of the door, Simon does not laugh. His silence is grave.

6

— CHARLIE TO SEE you.

— Tell him to give me five minutes. Yes?

Barbara says yes, reality says no. The office door kicks open, Charlie enters, bunch of flowers in one hand, cigar in the other — the flowers hit the desk and Charlie lands in the visitors' chair (fixed at a lower height than Howard's own, Howard paid a management consultant a £5,000 fee for this and other equally useful suggestions four years ago); he swivels, he leers.

— Welcome back laddie, here are some flowers, sorry I haven't been in to see you before now.

— It's good to be back. Thank you. They're very pretty.

— They're lilacs. A friend chose them.

Beneath his desk is an ugly brown bakelite vase. Howard half-fills it with washroom water, forces the lilacs in. He returns to the office and catches Charlie staring angrily at the photograph. Charlie looks away when he realises he is being observed. He glances at his watch. He picks at his fingernails with a pocket knife. Howard sits the vase on the window sill behind his chair. The flowers look good in the light. They also obscure the view of the wasteland. That is good too.

— Howard. I'm tired.

This is unexpected. An admission of weakness almost. And Charlie certainly does look tired. He sits heavy in the chair. The veins around his nose are like a colony of blood-glutted worms. His eyes are puffy and raw. He has kept his hair though and that is to his credit.

— Tired?

— Tired. Of working. Of scheming. Of doing the same things

289

every day. Of selling. Of buying. Of everything being predictable. I'm tired. Of this.

He waves a languid hand to indicate the factory they sit in. This is not what Howard had expected. Howard had expected the renewal of their battle. Howard had expected resistance, hostility. Howard had steeled himself to fight to the death to claim what is rightfully his.

— That's why I'm glad you're back. I'm done with the old ways. Howard . . . (Charlie sits forward in his chair. Charlie whispers.) . . . *I want out.*

Howard is confused. Howard had returned, in freedom, to sort things out. To restore the company. To avenge Mrs Buller. To defeat Charlie. He had not counted on a tired Charlie. On a yielding Charlie. He must be careful.

— Out of what?

— Out of the whole fucking thing. It never works, you and me. Now you're back you can take over. It's your turn again. Let me out of here. I'm sick of the commercial life.

— Are you running? You're not in trouble are you, Charlie? Something to do with an art scam perhaps?

Howard had not intended to spill the art club beans. Howard had intended to keep this bean to himself. But Charlie just looks unruffled.

— I'm not running from anything. Or maybe I am. Maybe I want to run from the whole fucking thing.

— I saw Leslie yesterday. He suggested that both of us should sell him our shares. It was, he said, his thought.

— Oh did he? How interesting. But forget about it. That was yesterday. Let's not look backwards.

Why not look backwards? If we don't know where we are then maybe at least we can find out how we got here.

Charlie gets up oh so wearily. Charlie walks around the room. Charlie circles the desk. Charlie grins. Howard shudders. Charlie has shiny teeth. Charlie is a predator. Charlie pushes the lilacs aside to look out at the sunless wasteland behind the factory and he nods at it as if he likes what he sees. Who would like a wasteland? Runaway killer convicts and broken bottles and shadows and mysterious parts of cars and scavenger animals and deflated footballs and ugly plants with straggly leaves

290

that twist and dispatch promiscuous pollen. Charlie used to hit him when they were small. Charlie became accustomed to cheat him as they grew older. Charlie is the ancient enemy. Time, if not Howard, appears to have defeated him.

Later, sneezing, in the attic, Howard finds his old table-tennis bat. It is inside a jumbled box of spoiled playthings: Scalectrix cars missing their connecting points and doctors' outfits stained with blood, a cracked Frisbee, four mangy shuttlecocks, a sad pair of deflated footballs. Howard's table-tennis bat is flat on one side and pimpled on the other. This disparity of surface is intended to confuse the opponent. It must have confused Vivien into assuming it too was damaged. Howard climbs back down to the corridor. This is the bat which has taken him almost every year to the semi-finals of the annual tournament, but never further. Howard essays a quick half-dozen imaginary backhand smashes. They feel good. He has always played better after a long lay-off. This might be his year.

Before Howard pays his nightly visit to Simon he stores the bat away in his attaché case, between the photograph and the Buller Report, beside four tuna fish sandwiches mouldering in clingfilm. Then he takes his place on the confessional cushion. He doesn't bother with introductions. He just starts talking as if they're resuming a conversation that had been momentarily interrupted. Simon will pick up the flow soon enough.

— We talked in my office for a bit and then, I don't know why, he said he had another appointment in a while, would I come into the boardroom and sit with him to talk until then? Weird, he wouldn't let me go. It was as if he didn't want to be by himself. Do you think his conscience has finally got to him? He's also got this thing with his watch he never used to have. He looks at his watch and then looks up as if he's calculating something. He did it a few times. Like a child trying to calculate how many seconds there are in an hour. It probably means nothing. He's always had a trick or two to distract the punters.

Charlie had energised his tired self with lines of cocaine that Howard had declined to share. Charlie's own portrait had been added to the gallery of stars adorning the boardroom wall. Made out of PVC, it was stuck gooily to a black background.

Not a sculpture, Charlie patiently explained, a bas-relief. Rainbow-coloured bubbles of plastic, they looked like a survivor of some awfully misguided manufacturing process, like the first stacking chairs they produced a generation ago. But, Howard had to admit, the artist (or engineer?) had captured something in his swirl of bulbous shapes, dynamic, greed in motion — do those splashes of flame signify eyes? — the juices of appetite corroding both the world and itself.

Charlie guzzled down the coke. He munched on his cigar. He expostulated cousinly love. Suddenly, solemnly — after another look at his watch — he got down to business.

— Look. Laddie. I know you've only been back a few days but tell me. What's the plan? Hands on? Hands off?

There have been times in Howard's life when everything appeared accidental — this wife, those children, that house, this family, that job, *my* history? This time was not one of them.

— I'm done with running. I want to make a go, a success at things.

Charlie nodded briskly, no attempt to make Howard's words sound ridiculous back to him, no attempt to claim absolute understanding of what was in his heart. Howard had expected Charlie to make these attempts.

— You want to make a success.

— Yes. I want to do something for the family.

— Big man. Excuse me, no sarcasm intended but it's more than that.

— I want to do the right thing by, by my father.

— Mop-haired little Danny. He's dead, Howard. It means nothing to him.

— For my son. For Simon and Sam and Dinah and all of us.

— Go on.

— This might sound absurd to you. I don't expect you to understand. I *want to make things right*.

Very gravely, maybe just a hint of humour or it could be malice in his bloodshot eyes, Charlie relented.

— OK. I get it.

— I want to establish myself here. I'm finding out what went wrong last time. I know you've been in charge while I've been away. I haven't come back to rock the boat, nothing like that.

Maybe I have actually. Maybe that's exactly what I'm here to do. This thing needs a lot of reorganisation. People fired. Departments disbanded. Concentrate on production again of profitable lines. We still manufacture the Unbreakable Comb, don't we? That shows some continuity. I want, this might sound silly, to do a *good job*.

Charlie nodded. Howard nodded. Together they nodded.

— That's good, Howard. I'm convinced. You'll want to buy my shares then.

— Excuse me?

— Think about it. It's the only way.

Charlie got to his feet. Animated by the cocaine, released by it being his turn to speak, he strode the room, his lips performed obscene acts with his cigar, he talked.

— This is how it has got to be. Leslie is right, it doesn't work with us. We're no good together. Like the song says, This town ain't big enough for the both of us. Howard, I envy you, you've got yourself a new start, everything's unwritten, kismet, you're your own man. Destiny is yours to build — You know what? We've got to get drunk together. Get shit-faced and roaring. Howard. You know. I've always had the impression that you think I don't like you.

Impression? *Impression*? Howard *knows* that Charlie doesn't, never did, never will, like him. It is their tradition. They've always loathed each other, tried to ruin each other, hit each other over the head, scare each other into falling into the river, will the other dead. Charlie must be up to something.

— What are you up to, Charlie?

— Up to? We're doing some exciting things, terrific things. I've hired some very talented Development boys, wild fucking imaginations. Whole new ranges planned. You'll love 'em. But fuck that. This is what I have to tell you, and I'm speaking to you hand on heart. I want out. I'm tired. I need a beach somewhere. Foreign sun. Like you, I expect, the things you did on your holiday. Saw some action, hmn? Howard the stud, look at you, years younger. What's the secret of eternal youth? I'll tell you the secret of eternal youth. Avoid responsibility. The mortgage, the kids, the missus, the decisions at work. Skip them all and you've got yourself a full head of hair. Mark my words, your hair is going to fall out again now you're back. No offence, but I thought I could already detect just a little hole in the middle . . . This is what I suggest: you buy me out. That's how it has to be. If your heart's so set on doing the right thing. Look. Even if I wanted to stay I'd be a thorn in your side, we'd end up fighting again, somehow, over something. It always seems to work like that. You've got to get me out of the picture. Buy my fucking shares. See it as the price of your return. Then the field is yours. Yes?

Charlie held his arms out wide, a lying fisherman, Christ on the cross, a boy about to execute an audacious dive.

— You want your old position back? It's yours. Snap your fingers, Howard Levy, managing director. King of the hill. Majority shareholder. The old office, the old routines, the old responsibilities. Me. I don't know. With the money I make from my shares I might even set up on my own, Levy Industrials? Levy Plastics? If I had children it would be Levy and Son Manufacturing. Don't worry. I'm joking of course. The last thing I want is to stay in this business. I'm done. We'll work through Katz, we both trust him don't we? Make me an offer.

For a moment Howard swayed first one way then the other. Yes, Charlie was right. That was the way it had to be. There was room for only one of them, history had shown them that much at least, and here was Howard, returning, lusty, vital. Time for old man Charlie to take himself a rest, a retirement home for has-been conmen whose moves have got rusty and whose blood has flowed cold. Or what if Charlie was right but it should be Howard who should step away? Levy and Son, set up on his own with an empty office waiting for invisible Simon. Build junior a suite with reinforced walls and a huge security door to hide behind, multiple locks, that should excite him. No, this is what they have, LPC, their world has been established before them, by dead Levy men with mysterious intentions and adulterous habits.

— I don't have the money.

— Then raise it for fuck's sake. Look. Here. Treat yourself. Have a line of coke, that'll loosen you up.

Howard jerks back. Howard bangs the back of his head twice

against the landing wall. He brushes some hairs off his pinstripe suit.

— God! It's just hit me. Embarrassing, I should've realised earlier. Shit. Sorry but I've just got it. He's still pretty good, Charlie. He's made a life out of being an operator and I'm the one who's rusty and I forgot the golden rule. I learned it and relearned it years ago. Paid enough for the privilege. When Charlie says he wants something then you have to look at the opposite to find out what it is he really wants. The fucker. Excuse me. Trying to make me bankrupt myself buying his shares. And then what? The company must be about to fold. I've seen the surface of things. Underneath it must be worse. That's the only explanation. It was in a bad enough state before. Thank God I didn't commit myself to anything. I knew that much at least. I played along. Sure. Yeah, Howard the sucker, nothing changes. Pills? I'll buy them, how much? Two shillings to look up Millie's skirt? You sure she doesn't want more? Buy your shares in the company? Of of course old fellow, delighted I'm sure . . . But he's going to get a shock. Buy him out? No way. I tell you what I'm going to do. I'm going to sell *my* shares to *him*. I don't know how to work it but that's what I'm going to do. He's good, he's still good, he nearly had me, trying to stiff me. What do you think about that? It's the table-tennis tournament tomorrow. That's got nothing to do with anything but I've never won it and I'd like to this year, my last act as company man. I'm going to assert myself. Fuck Charlie. I'm going to sort him out, I tell you. I'm selling, not him, and at *my* price. Then maybe I will set up on my own. Do you think there's much of a market in self-decorating candles? Levy and Son. Or Levy and Daughter, I'm sure I could persuade Ruthie to go into business with me, it's not like she's doing anything else with her life. How was your dinner? Have you finished with your tray?

7

HOWARD'S HAND, CLEVER little animal, completes its rummaging. A squirrel of a hand which has just discovered its nasty rotten little nut. There is a space there, in the middle, a small space only but a space all the same. And bigger than the last time he checked.

He might tell Simon that he visited the factory floor out of a sense of duty and also as a way to ground himself. This is the kind of talk that Howard is sure Simon likes. But really he does it in the headrush of an inspiration. With Danny's sketch in his hand, and with tennis shoes on his feet, he flies out of the office area and down onto the shop floor. The air tastes reliably sour as he walks past the desultory workers. There are fewer machines than the last time he was here, on his midnight visit with half-forgotten Carol. The machine absences are readable in the lighter-coloured patches on the floor, indentations of pigeon-toed metal feet, perfect clean squares, surrounded by drips of oil.

The men in overalls and vests and sweat and oil look up at him as he goes past. They don't even bother to pretend to be working harder than they are and that's another bad sign, of morale and of their opinion of him. He walks quietly through to the far corner of the shop floor. Howard reaches Museum Corner. Five squat machines, starved of electricity and love, soiled by time, wait for him there.

Several times attempts have been made to uncover the secrets of Danny's last machines. After his death by fire Sam and Freda interrogated for days the Slav Ivan who had built these girl prototypes under their inventor's supervision. Ivan knew nothing, hardly spoke English. His muscles were good, his curiosity absent. He shook his head often, stared at his interrogators through worried little blue eyes, he knew only one thing, his conviction that his new bosses were accusing him of stealing. Sam put the company's top engineers on the case. Freda brought in consultants from outside. They all dirtied their hands and scratched their chins (transferring Danny's original grease to their smooth white cologned faces) and talked enthusiastically about piston screws and movable die plates and runner systems. These machines were, they decided (whisper the words, do not offend the family priests) *primitive, simplistic*. Each at best performed a single function that other machines knew how to do better. The engineers were sacked, the consultants paid off. The machines were left to rust, their girlish names hidden behind years of grease and neglect.

Howard, inspired Howard, regards them one by one, compares their details with the candle manufacturing machine sketched down out of his father's imagination. This rheostat, is this the one marked *l*? Or that cooling feed cylinder, isn't it, if he just turns the sketch around forty-five degrees, isn't that *p*? No. No matter what angles he chooses, none of the machines corresponds to the sheet in his hand. Again he has shown himself that he is not his father's son. So much for inspiration.

If he can't connect directly with the past then at least he can keep it clean for someone else. From the maintenance shelves Howard picks up a tin of grease and one of the least dirty rags. He starts with the brass plate of the first machine. Digs a corner of the rag into the tub of grease, smears clean circles through the dirt, letters become apparent, fluently carved with an italic flourish in brass, then a name, *Rosie*. He moves on to the next, leaving Rosie, her name at least prettily shining.

Howard works on ignored. The brass nameplates shine. Rosie. Bertha. Ruth. Agnes. Christine. Different shapes, these machine girls, some squat, a lever for a nose, screwholes for eyes, an automatic instrument panel triangle where pubic hair belongs, others narrow and angular with pairs of red buttons for breasts.

And when he is nearly done he notices that he is alone. The workers are gone, the machines silent. Howard hurriedly digs out the semi-circles of oil that have made an *o* of the *e* in

Christine. He leaves the rags and the tubs of grease and wax where they are. He washes his hands and forearms under the cold tap in the corner. He runs up to his office to fetch his case. (Remembering to pack the photo, his good luck charm, inside.) He's probably already late for the opening games in the tournament.

They are all there in the R 'n' R room. Everyone who works for the company and a few who were sacked along the way. The annual tournament. The last tradition, which began with Sam who loved competition and sport. Two table-tennis tables wait in the middle of the room. Around the side workers and management drink from cans of beer and lean against pock-marked walls. Breathless Howard is drawn against Arnie in the first round and that's an easy game. Arnie (with a red bandanna tied around his ginger head) has developed a new kind of spin and all Howard has to do is to bounce the ball over to him and watch his opponent make mistakes. 21–11.

Charlie plays against Nikki who is dressed becomingly, if over-elaborately, in sweat pants and leotard and sweat bands on forehead and wrists. Charlie, in his regular clothes, but with trainers instead of brogues, takes it easy, loops the ball over the net, gallantly looks as if he is trying harder than he is (which infuriates Nikki). He keeps the score close. Whenever she threatens to take the lead, he sends the ball back over with a nasty edge of spin, confounding her completely. 21–17.

Howard looks for his attaché case, which he has foolishly left out in the open, where the workers are drinking and flirting. It's not where he left it. Is that victorious, sweating Charlie skulking out of the room? What is it he's carrying in his hands, hidden by the sly bulk of his body . . .?

SEVEN

TWO PER CENT

Fly me, sang the Hoboken serenader, to the moon. When we are flown — or no, when we *fly ourselves* to the moon, and on, to the stars, those amiable worlds that we see twinkling in our winter skies at night, what will we wear for wings? They shall not be of wood, nor feathers, poor Icarus. Bronze and silver and gold are too primitive, inflexible, for us. While iron and steel are quite beneath us, materials for a dirtier, less elegant age. We shall surround ourselves with mobile luxury cocktail bars and swimming pools and libraries, as we make with speedy efficiency for the stars. We will enjoy our travels and all our new discoveries with ample good nature. Because our wings will be made of plastic.

H. C. Willi, *Some Tentative Considerations Vis-A-Vis The Future*, epilogue to *Century of Plastic: A Celebration.*

I

HE IS ALMOST there. The banana boat to La Paz. Conchita the harbour-master's daughter. King Charlie chewing on coca leaves in his swaying jungle hammock, his royal skin cooled by the palm fronds waved by Conchita's naked sisters, naughty Chiquita and the darker-skinned Ramona and there, the youngest one, might prove to be the pick of the bunch, give her a couple of years for those buds to develop into breasts, saucy-eyed Lolita. But he is not quite there yet. So says Leslie, his accidental guide through time. With an eye on the law and a hand to his dripping nose and the other pressed to his own scared neck, Leslie counsels patience. Seventy-six per cent and he can do what he wants. Shut the factory down, blow it to the skies, sell to Wiener, to Chatterjee, throw the last, apocalyptic party. *Try to close deal now*, warns moist Leslie, *excuse me, my nose, this weather* — (a sudden nasal explosion which makes Charlie wince; his own poor nose would collapse at that kind of treatment) — *that's better, try to sell now and we'll have a war on our hands*. And then, seeing the dangerous look in Charlie's eyes, *Wait, please, I beg, for Howard to yield or Dinah to, God forbid* . . . (The prospect is too sad, he cannot give it words, it expresses itself in tears poured into the handkerchief conveniently already held nose-high.)

Charlie gets up with a framed photograph just purloined snugly hidden in the waistband of his trousers. He is shocked by the treatment it has received since he last had possession of it. It has been ripped and battered and creased and generally dishonoured by sweat, dirt, destruction, age. It has found its home now. He will look after it. Unlike the simpleton sandwich collector. Charlie passes around fresh beers to raucous swearing

301

workers. Allows himself to feel a little sentimental. He was thirteen when he entered the competition for the first time. Lawrence Levy, heir of the recently dead Sam, had been allowed, droit de seigneur junior, to win a couple of games until he came up against Avi the foreman in the quarter finals. Avi was his ping-pong teacher, taught Howard too, in school holidays, rescued the pair of them from murder attempts on the other, rescued them from the boredom of crayons on ancient blueprints. Avi was fired years ago, by Freda, when she found him doing what everybody else knew he did every Friday lunchtime, stopped the machines to sell costume jewellery to the workers. He returns though each year for the table-tennis competition, white-haired now, raggedy clothes, arthritically stooped, beer madness in his little blue eyes.

Charlie will be good to these men. Their pay-offs will be handsome, redundancy cheques to piss away in beer or lose on the horses or put a deposit down on a briefly tidy house for their enormous raggle-taggle families. Charlie sits down again, beside Mark. They watch the games together. Avi's on the S table, sad wisecracks puffed like pipe smoke from his mouth as he gives Angus a lesson in guile and spin. You can tell a man by the way he plays ping-pong. The killer instinct or the lack of it. The losers and pussies who serve into the net when match point is against them. The smooth bastards with their stylish smashes and nothing else. On the D table Howard, grey suit, white tennis shoes, resolute defence, bald spot shining, wears down his opponent with steady backhand returns. Howard, you fucker, twiddling your isolated bat, waiting for your opponent to serve.

— Who's that he's playing? Bill?

— William. Crazie Howie is a better player than he used to be.

— You think?

Silly question. Mark never lies. Often doesn't talk at all but when he does he's always saying what he thinks is true. Crazie Howie, do you know that's what they call you? With your jesuitical eyes and your leading questions and the way your hands twist at the wrists and open and close over nothing whenever you're nervous? Your prim little attaché case, what

302

did you think those locks were for? Carrying that photograph around each day, next to an increasing pile of tuna fish sandwiches. Did you not know that Charlie would come snooping? Did you not know that he would steal your lucky charm?

Stupid Howard. Crazie Howie. Trying to fight history, trying to breathe his own soiled air into a corporate corpse. Proving himself in his second reign to be even worse than he was in his first. Charlie had known how it would be. He had seen it all. Put Howard in charge and wait for it to get on top of him until he begs to give it up. Then add the enticement of making Howard think he was putting one over on Charlie. Tomorrow Charlie meets Chatterjee. The final deadline. There will be papers to sign. Howard or at least the company will have to be in his pocket. Howard has got wind of the art deal, Charlie's natty way of manufacturing extra cash to buy other people's shares with. This is not a disappointment. Makes a game of it.

— When's my next game?

— You got time.

Charlie slips away. He has heard the call of a heroising pinch of cocaine.

A burning in his nostrils (what is a septum anyway?), a sudden drip-back of bitter aftertaste, a surge through his muscles and brain of narrow. Pure. Unstoppable. *Purpose.*

Power. Strides back into the room. Big as a king. Yeah, you too Gazza, how about it? The sounds around him are reduced to a clean protective wall. Gets the nod from cold-eyed Nikki, the tournament organiser, takes his place at Table D. Warm-up. Doesn't reveal his best moves. Returns the ball same angle it comes, a dash of top-spin, a kiss of side. Loses the knock-up for serve. Do what you can, boy, because there's nothing you can do. Who's he playing? Who cares? The marketing brat Gary, they call him Gazza, with a baseball hat turned the wrong way. Call that a serve? Piddly little slice, top-spin it back, long, set up for the smash, BOP. Thank you very much. Yeah rhythm, rat-a-tat, backhand, backhand, backhand, forehand smash, return *that*. Oh is that the score, lucky me, I think not. Backhand serve, forehand serve, spin on it crazie as Howie, straight backhand when the kid's expecting the spin, loops off his startled bat straight into the net. Adore me.

The king is in his palace. Everything is shimmering, translucent plastic. Beautiful, disposable. Invisible plastic walls cut into the earth, touch the stars. Surrounding him are wind-up toys with painted faces, rattling, shaking, drinking beer from plastic cans, mechanical arms up, mechanical arms down, pick them up, inspect their revolving bottoms, made in Hong Kong, made in China, made in fucking West London, those days are gone, excuse me I wasn't ready, never mind, fuck it, your point, whose serve? what's the score? yeah? really? that close? Suck on this, *zip*, serve goes through — it touched the net? oh, did it? it won't this time. Power.

Charlie wipes the sweat off the handle of his bat as he settles back beside Mark. Mark gets up, his game now, bound to win, religious urgency, athleticism.

— Charlie.

Why, hello. It's Howard sitting where Mark used to be. Howard, Daddy's little loser, he's a plastic marionette, knees are right angles, strings attached, you can see where they go, his hands rest on narrow lap, ready to be lifted into a dance for children to laugh at.

— We're playing each other in the semis.

Oh my. Almost sounds like a challenge.

— You against me, and in the other Avi against the winner of this game, probably your driver I expect. He looks like a sportsman.

— He used to be a footballer.

— We should talk. Weren't we supposed to get drunk together? The company's got problems. We could maybe sort them out if we got together.

— Tried to kill a referee, got himself banned for life.

— Look. I know what you're up to.

— Maybe not kill him exactly, more a sort of execution by GBH.

— Charlie. *I know what you're up to and I can stop you.*

He's bluffing. Never knew he had it in him. How charming.

— Yeah. Probably. He says, though, it was worth it, the referee. It was a youth team cup game, crap referee, kept making bad decisions, but the thing was, most of them were in favour of Mark's side. That's not the point is his angle, the point is the decisions were *wrong*. Mark's a moralist, you know. Ref sent off his third player, I think we're somewhere in the middle of the second half, should know, 've heard this story enough times, a bundle of kids standing around the ref, too baffled and angry to even shout at him, just standing there, staring, disconsolate, you know, it's like God's gone mad so what can you do? I'll tell you. Mark just walks through the bundle, that quiet way he has, sinister, pushes a couple of his mates aside, gets to the ref who's fiddling with his little book, *pages* covered with names, and he looks up to see Mark and Mark I think headbutts him first, poor stupid little bloke goes down, and Mark just keeps on kicking him until he's not moving any more and that's when our lad lets himself be led aside. They finish out the game, linesman takes over the whistle (first thing to do is send off Mark), Mark's team wins, like it was going to, two clear goals, he'd scored one of them, the team gets through to the final, might even have won it I don't know. He spends some time inside, gets himself banned from playing organised football anywhere in the world for life.

There is a space around Charlie and Howard, their own private patch, cigarette lino floor, the two plastic chairs they sit on, backs against the wall, seats angled a little away from each other. The ping-pong tournament is traditionally misrule day, when the workers get drunk and the bosses get mocked but still, even so, respect has granted them an executive airspace. Or maybe it's the smell of something dirty that's produced when the two of them combine and which neither has got the nose to smell.

Sometimes Charlie longs for a controlling power equal to his to hammer its siege phallus against the plastic portcullis of his plastic palace. So tempting to shut eyes, drift into swimming pool laziness, the barbarians at the gate, surrender. He wishes Howard's strings and clockwork mechanism were not quite so clearly visible. He wishes Howard were smarter. Maybe he should just try the direct approach. Stop beating about the fucking minge.

Oh the delight of it, to risk it all, the grand coup. Of course he can beat Howard with trickery, but Charlie's life must have a grander meaning than that, the proof over and over again shown how to use his cousin's stupid nature against him — but get this, to do it straight, in public, in glory.

— I want to buy your shares.

— I know how Mrs Buller died.

So the fuck what?

— Indeed.

— I could get you sacked. The art club thing.

Perhaps so. But Charlie doesn't believe it all the same. It's not the way Howard does things. Let's spice this up a little. Let's get the juices really going. Let's put it in the hands of fate and skill. Thrilled Charlie whispers.

— Howard. An idea. You were going to try to sell your shares to me, admit it. And I'd have bought them and in a while you'd have found out why and hated me and hated yourself for being so stupid. But let's do it this way. Ping-pong. You and me. I beat you and you sell me your shares. (A quick calculation: at the market rate there is just about enough cash in the Classic Art Club™ fund. Fortunately, Howard's stewardship has driven down the value of shares even further than before.) You beat me and, fuck it, I don't know, I sell you mine? I retire? Whatever you want. How about it?

It's a suggestion that appeals to them both. A pointless, stupid gamble to settle things. Maybe they are like each other after all.

— If you like. Why not. But why bother with money? Let's do it properly. Winner takes all.

Oh like it. Never knew the lad had it in him.

— Winner takes all.

They shake hands on the deal. Baffled, loving Howard is staring him in the eyes. Charlie has to look away, it's so nauseatingly meaningful.

— We need to be friends, Charlie.

Said sadly, a warning of sorts. Pathetic.

— Course we do, old son. It's everything I've ever wanted. Excuse me for a moment, will you, got to shake the snake.

Man, this is deadly, like their old-time jousts on bicycles in the wasteland. Everyone's watching. The room is nearly silent. The big cheeses are at the centre. The bosses are fighting a duel. Howard's serve looks innocuous but has got a nasty *tone* to it, stays low, sidles to the net, rolls over and tries to play dead. Charlie lunges forward to lift it back and a passing shot zips past his elbow. Love-One. Um-*um*.

Howard returns well. Dead bat against anything aggressive. Everything Charlie fires comes back. Smashes, corkscrewing lobs, flat top-spinned forehands to the far corners of the table that *nobody* ever returns. Charlie changes strategy. Fight water with water, he sends over looping daffodil spins, the enemy doesn't like to commit himself, doesn't usually go for the killer shot. Agonies of attrition, what are the crowd doing? neither Charlie nor, he is sure, Howard knows. The world is a ping-pong table. No one else but Charlie and Howard and the ball and bats and table that connect them has, could have, ever existed. Rallies go on for ever, the ball bouncing on the same spots each time, short corner to short corner, each of them waiting for the other to make his move, neither does, same backhand each time, just on the edge of the table, the ball moves with a constant speed, someone coughs, Charlie smashes, break this fucking rhythm, the ball misses the table. Three-Nine.

Loop, loop, long shot, soft shot, *smash!* Yeah, that got him. Backhand, backhand, backhand, forehand, backhand, and then a lofty controlled backhand smash, a conceited high-arcing power shot, bounces off Howard's bat, bounces off Howard's nose, bounces up the table, dribbles over the net. Seven-Ten.

— Sorry mate. Noses not allowed.

A weak joke but everyone laughs all the same, gets the audience on the side of the king, we need more noise here. Now we're talking. Rap. Tap. Smash. Rat-a-tat. Loop and smash. Thirteen apiece. By his game you shall know him. So what has Charlie learned so far about Howard? — Other than he's a lucky fucker: a rare attack, a long opposite-corner half-smash slices down against the tip of the white corner on the very edge of the table. Answer that with a regal forehand smash that sends Howard scurrying to the far wall to retrieve the ball and gets the audience clapping. Fourteen-all. — He already knew that Howard retreats under pressure, that he's too essentially passive to make the bold dangerous move — Howard's second-best serve, the slicing backhand dink, has become predictable, Charlie sets himself up for an easy smash: Sixteen-Fourteen. — What he never realised before now, and how was he to know, given Howard's runaway history, is that the stupid fucker never gives up. Keeps on. Chases after everything, takes a pratfall, lands on his arse as he sends a high loop back, which Charlie, distracted by his opponent's inadvertent clowning, feverish with the blood-scent of victory, smashes into the net. Sixteen-all.

Seventeen-sixteen with an unreturnable dinky serve straight out of Howie's playbook. Eighteen-sixteen when Howard hits a hard skimmer straight into the net. (*Hold on Lolita, Daddy's coming!*) Eighteen-seventeen when desperate Howard essays a smash and, lucky fucking fucker, it comes off. Your serve, you fucker, the world's disappeared again, it's just you and me.

Howard catches the ball, he makes it roll and stop ready to be served in the steady flattening saucer of his left hand. His eyes are looking for Charlie's. Charlie doesn't fall for that. He learned that lesson painfully, ten years old, pulled out of the audience (his hand had been higher than anyone else's) to play the stooge at a magic act. Watch the eyes or the seductive other hand and you end up with a pink balloon twisted to look like a dog, people laughing at you and a patronising clap back to your seat. Watch what he's doing, really doing, and you've bought yourself a piece of the action for free. Charlie keeps his eyes on Howard's bat hand, swooping down towards the ball.

— Mr Levy?
— Fuck off. — Fuck off.
Times two, simultaneous, both Mr Levys curse the approaching world; the serve skims over, Charlie hits it back into the top of the net, it wants to reach over, it almost does, it falls back rolling towards him. Eighteen-all.

— Mr Levy. Mr *Howard*.
The world cannot be pushed aside so easily. It encroaches in whispers and beer and rude colour. Valerie, pregnant again, floral dress with a big pink bow over her bust like we'll find a fucking box of melting chocolates concealed down there, steps forward, lifts a dewlappy arm to Howard's shoulder. He crou-

ches down so an ear will be at the same level as her whispering mouth. Patient Charlie scans the room. He can see every slow-motion detail, the broken-veined blemishes on every face, each surprising flash of beauty, a girl's hipbone when she turns to ask for a light, a man's bicep with its rough tattoos of a red heart and green flag, the parabola of beer from the shaken-up can which someone had just forced a screwdriver through.

The crowd don't like this interruption either. They want to see one of their bosses kill the other. They might even have a preference over which one should die. Valerie whispers. Howard nods. People shout, *GET on with IT!* A slow handclap builds. Charlie tries to keep concentration. He bounces a ping-pong ball up and down on his bat. Howard is listening to Valerie but looking at Charlie, sad? hopeful? Masked. Howard lays his bat down.

— Charlie. I . . .

Not masked. Something plaintive in Howard's face. A naked need for help which Charlie can turn to his advantage.

— Troubles, Howard? Fuck it. Your serve.

Howard looks at his bat, flinches toward it, shakes his head.

— I can't, I have to, sorry.

About turn. Howard is suddenly gone. His bald spot the last thing they see of him, the absence that precedes absence. Charlie is through to the final and Charlie owns the company and Charlie feels cheated and tricked. The game is forfeit. Is that it? So easy? Company control passed over, painless, moneyless, bloodless? Maybe the game is void. There's no rule book to consult. Maybe they have to start all over again, except now Howard knows what Charlie really wants. That wouldn't be fair. It was Howard who left the game. It must be forfeit. Charlie returns to his seat, watches Mark and Avi start to play their game. An intriguing contest of power against guile which —

Charlie *knows*. He jumps up, rushes out of the R 'n' R room — a quick series of comedy routines as employees pretend to be getting out of his way when really they're blocking it — down the corridor, up the stairs, two, three at a time, cocaine athleticism, into his office, makes the telephone call, a number he knows off by heart. It rings for ever, finally answered. A female.

— Scott Ward please. Make it quick.

— He's not in his office.

— What?

— I could page him for you if you liked. If it's important.

— Page who?

— Doctor Ward. Is it important?

Start again. Resist the urge to curse and shout and crush.

— My fault. Sorry. I want the Robert Falcon Scott Ward. Now.

— Oh, ha ha. I thought —

— I know. Never mind. Please.

Another eternity marked by the torture trills of the unanswered phone. An Irish woman finally picks it up.

— Scott Ward.

— Charlie Levy. Dinah Levy's nephew. I'm calling about my aunt. Is there any . . .?

A silence. Heavy. A new voice comes on, equally Irish, less womanly.

— Mr Levy?

— Yes?

— Are you sitting down?

— What do you mean, am I sitting down? Are you wearing lacy knickers? Who the fuck cares. I'm calling about my aunt.

— It might be better if you were sitting down.

— All right. I'm sitting. I'm sitting. (Charlie, who in fact had been sitting, stands up now in his frustration.) Get on with it.

— I'm sorry to have to be the one to break this to you but your aunt passed on earlier today. She suffered, I am glad to say, no pain.

Another pinch of cocaine (no time for mirror and banknote-straw, this one snorted from the snuff spoon of his flat left hand), then back out again, a magical descent down the stairs, feet dancing hardly touching the steps, along the corridor, no, stop, the screech of rubber soles braking against corridor floor, no time to fetch Mark, Charlie'd have to explain things, the inertia of people would suck him in there for ever — Charlie doubles back, through reception lobby, shoulder charge against door, the slap of outside air, into the car park, he'll drive himself for once, almost to the car, fuck it, hands pat pockets, you arsehole Charlie, no keys. Back into the building, towards the R 'n' R room, away from it, towards it, away from it, it'll be quicker to get the spare set, up the stairs again, two at a time, can't do any more than that, he's out of breath already, back into office, top drawer no, just his drugs stash there and seven paper clips and an ancient self-correcting typewriter ribbon, next one, documents, fuck the documents, make them fly into the air, a blizzard of paper, got it, the spare set on a football keyring, out of his office again, down the stairs, really out of breath now, jog through the reception lobby, out into the car park, open the Scimitar door, wrong side, he's beginning to panic, hold it down, his chest hurts and so does his head and his heart's going wild, stay in control, that's the passenger side, run around the car, driver's door, it's locked, fuck it, which key? unlock door, slide into the seat, no time to recover breath or practise with the gears or rub the knee stinging from its collision into the steering wheel column, start her up, nice homecoming roar, she's pleased to have her Uncle Charlie at the controls again, out out out.

2

THE FUCKING HOSPITAL. Screwing out its little bit extra by making people pay to use the underground car park. Reversing has never been Charlie's game. He grazes a concrete pillar, scratching a grey line through the racing green of the side of his car, curses, lurches forward, stalls, fires the ignition again, reverses, digs the scratch deeper, and leaves a cigar butt smouldering inside his bruised car parked across two spaces reserved for hospital staff.

Automatic doors, corridors littered with the tin can wreckage of broken old people shivering in wheelchairs. An aquarium green with mossy water and brown with sluggish fish. The lift that takes an age to carry Charlie and three giggling, thick-ankled nurses up to the third floor. It is imperative that he get there in time. The final effects. The last two per cent. Charlie is not so naïve to believe that he can hold Howard to their deal. Howard will never agree to sell to Chatterjee. Not now. Not Howard the orphan. He might though to Wiener. Could Charlie sucker him that way? Like he suckered Leslie. (And he mustn't forget to punish Leslie for his presumption in trying to buy Howard's shares.) Play the family card, we must pledge ourselves to history, this is how we can keep the business going, sell to a man who knew our fathers. And the site? Oh worthless my lad. All those pollutants that have seeped through over the years. Impossible to build there, don't worry about it, I'll dispose of same for a consideration, I know a fellow on the Council . . . but the business we'll split it three ways, you, me and Leslie, what do you say? You tell me we can't build anything there? Certainly not. No houses there or shops. But what about a car park? Well, yes, a car park possibly, but who would want a car park in the middle of nowhere just off the Great Western Road? It's not like there's any new development planned in the neighbourhood, nothing like that . . . Can he keep it secret? It is tantalising. Lift doors open, wriggle past a trolley containing some wizened fruit tied to drips. Turn left, turn right, brush past medical staff eager for conversation, *Not now son, it's a battle to the death here*. Into the Robert Falcon Scott Ward.

Four beds and one of them, Dinah's, is freshly made. Hospital tidiness, Death's fastidious friend, has smoothed out any sheet-crease memory of the recently dead. Starched hospital sheets. Blue hospital blanket. No trace of his aunt's head on the plumped-up pillow. The three old vulturesses lie on board their beds, each of them staring at a dead TV screen.

— It's a sign of respect. Sweet.

A pretty nurse by his side. Blue eyes. Freckled face. Name badge says Mary McDonnell Staff Nurse.

— That's how we found them earlier. After lunch. Your mother . . .?

— Aunt.

— Your aunt had passed on. Such a peaceful look on her pretty face. You might like to see her? Many find it helpful.

A decisive shake of his head. He certainly does not want to see the husk of Dinah, no matter how peaceful the look on her dead face.

— These dear friends of hers switched off their tellies straight away. It's their little way of honouring the departed.

— Very touching. I . . .

I what? I would like to read every word of her case notes while you sit on my face, and every now and again I will brush away that strand of pretty brown hair that drifts down over your perfect neck? I would like to fuck you? Death is making Charlie horny. He needs to find Howard. He needs to make sure his cousin (Eighteen-all) isn't getting a jump on him. He needs to stay on top of things. He needs to ram his cock home into Staff Nurse Mary McDonnell until the two of them are screaming.

— You poor man. Would you like a cup of tea?

Charlie nods. Gulps back the pornography like tears.

— Please. Come with me.

Her fingers on his arm gently leading him into the staff room are an exquisite torture. Dare he? Can he stop himself?

Staff Nurse Mary McDonnell pours grieving Charlie Levy a generous mug of sugared milky tea. He lifts the unsteady mug to his face and takes great draughts from it as she clasps his free hand between hers, pulls it down to the bony freckled comfort of her clinically covered lap. He tries to keep his hand still. He cannot. He stares at the floor with — he hopes — bitter sadness while he penetrates her fingers with his. His thumb has discovered a blue hole in her uniform. It quietly makes it larger.

Charlie Levy gives himself up to the need, and so does Staff Nurse Mary McDonnell. Charlie doesn't wonder at it for long — maybe this is what she became a nurse for, maybe she is as turned on by sorrow as he is by death, maybe they are each other's one true love — as her uniform rides up her thighs, as he pulls down her knickers and tights and she pulls him down to her on the floor, as her whispers to him alternate between chiding him for his wickedness and comforting him on his sorrow, as her fingers get trapped within his pompadour of hair and his introduce themselves to the inside of her arse, as they fuck each other in a tatty room where the unlocked door rattles gently in the tranquil hospital breeze, his mind and, he is sure, hers, finds a kind of stillness and peace.

3

CHARLIE LEVY, GANGSTER Charlie, lover Charlie, zips up his fly, remoulds the photograph around the comforting hill of his belly, and walks on down the corridor. Staff Nurse Mary McDonnell has stayed behind in the staff room to reright chairs and gather mugs from the hiding places they had rolled into. The corridor seems deserted. The nurses' station hub abandoned. Charlie leans upon the nurses' station wall, examines blinking lights on telephone consoles, misspelled messages on medical information forms. He turns and leans against the wall. He inspects each corridor in turn. He needs to find Howard, he needs to introduce himself into the story of the aftermath of Dinah's death. It is awfully sad, of course it is, but that was no way for a dog to live and anyway he is glad that there is no one left to call him Lawrence.

— Charlie!

All things come to he who fucks them. There's Howard approaching down the corridor to his right (nasty smell down that one, and darker too — Charlie suspects experimentation zones, incinerators, dark practices). Howard is pale with sadness. He has been crying. He holds a supermarket carrier bag tight to his chest. His table-tennis shoes look forlorn beneath his company suit. He carries himself with dignity. Sorrow surrounds him, almost, it has to be said, becomes him. Charlie's busy heart goes out to him.

— Howard. I —

— Where have you been? I've been looking, calling. The factory. They said you'd disappeared.

— I was —

— There was something I, I . . . *needed* you.

Awful admission for Howard to make. Charlie wraps consoling arms around his cousin.

— I'm here now. Charlie's here now. Poor Dinah. Poor Howard. Spill your heart.

Howard shrugs him off.

— That's not. Not. No. There was. Shit.

The man is crumpling. Normally that would be a delight to see.

— Come along. Let's have a sit down. A cup of tea. I know a place we can go.

Back along the corridor, into the staff room where Staff Nurse Mary McDonnell repins the white curve of cardboard atop her tight bun of hair and daubs twin clouds of dreamy blue shadow above her sweet blue eyes. Eyes which suddenly turn wary at the sight of the men. Her thin lips press protectively shut. *Him too?* Charlie reads her mind, *You bastard. I'm not a machine.* Charlie slaps her narrow arse to set her at her ease.

— Don't worry, gorgeous. This is a job for Charlie. My cousin here, Howard. Howard, say hello to Staff Nurse Mary McDonnell. I thought — Excuse us, love. We've got some talking to do. I thought? I wondered?

— Of course. (Brisk again; a lava of pleasure rolls half-hidden beneath her professional prim.) You know where the tea is.

She leaves behind a smile that would brighten up the soul of anyone less sour than Howard.

— Sit. I'll get the tea on. Talk.

Howard talks. He tells Charlie about arriving at the hospital. He tells Charlie (impatient Charlie but magnanimous Charlie, considerate of Howard's sorrow, hides his frustration at the slowness of the story by taking pains to spoon-squeeze every drop of tea from Howard's teabag) about his dread as he approached the Scott Ward.

— I walked as slow as I've ever walked in my life.

The telephone message from the hospital had been garbled along the way. The hospital to Valerie to Howard. Valerie's Chinese whisper to Howard had told him his mother had fastened. *Fastened?* What did that mean? Starved herself? Been attached to some grim new apparatus? Sucked herself on to her disease like a baby to a nipple? Sickening realisation that what-

316

ever the code the message was clear. Dinah's death, missed. Dinah's body, waiting.

— A nurse told me she'd passed on. It might have been that nurse we met before. (Accusing look at Charlie: *You.*) No. Couldn't have been. Maybe that's what they teach them to say. Passed on, fasten. What does *passed on* mean? Passed on to where? This tea's too strong. Can't drink it. They asked if I wanted to see the body. I said yes because that was the easiest thing to see. It looked well-made. It wasn't her. She wasn't there. She's gone. Not passed on. A puff of smoke. Evaporates. No more.

Howard's elbows dig unreasonably into Howard's thighs. His left foot is tapping out an unsteady rhythm. His hands are a cup for his chin. His mug of tea sits rejected on the arm of his chair. He stares at the wall opposite, a poster facing him there, tourist enchantments, *The Magic of Andalusia*, different frames each showing a different magical attribute, and he looks at it as if he is trying to remember where he saw it before. He keeps talking.

— I wasn't in my right mind. I'm still not. Frantic. Like speeding. Remember when we used to do amphetamines? You ripped me off always. Paid for your own with some of the money I paid for mine. Knew it then. I had been expecting it, of course I had, and I was hoping for it too, it wasn't right for her to be like that, we all knew, even, even, even *Leslie* knew that. But when it comes how do you react? There's no precedent. You sort of wish someone would make it like a film. Take you into a room. Cover everything in velvet. Light some black candles. A woman without a name walks through under a veil. Shadows. Instead I get this.

He looks around the room. Blames Charlie for the sink full of unwashed tea mugs, blames him for the tourist poster opposite whose twin he still can't place, blames him for the neon strip lights that make everything look as if it were about to break.

— Everything was confusing. Still is. Sort of, I don't know if this makes sense, *unanswerable*. Does that make sense?

Charlie pats his cousin on the knee. The right words from him now will put Howard for ever in his debt. Has to be acted on. Also — a less powerful impulse but Charlie feels it there all the same, flexing unused muscles — he would quite like to make Howard feel better.

— She had tried to say something to me before. Before. When she wasn't quite in the coma. I couldn't *hear* it. Maybe she was telling me what she was going to do, had already done.

The wrong words will ruin it. Or maybe it's just the fact of words, any words will get the job done.

— Don't worry about it. It's over. You've been a good son. You still are.

Risky. Howard glares at him. Continue. Can't bottle out now.

— It's how it has to be. Parents die. Children mourn. Better that than the other way around. It might — excuse me if this sounds presumptuous — be good for you to cry.

Agony uncle Charlie, doing his best to perform his unfamiliar role of Howard's comforter. Howard bursts out laughing. Ugly sight. Even uglier sound. Looks like a hyena. Sounds like a chicken.

— You don't think . . .? Really. Is that . . .? I'm sorry, feel free to correct me if I'm wrong but are you going to tell me now that time is the great healer?

— Don't get snotty. I'm doing my best.

The family together in strife.

— Charlie. Believe me. I know you're trying to be nice, but do you honestly think if I wanted sympathy I'd come to you?

Charlie shrugs. Contemplates the damage that a single punch from a single fist would wreak on Howard's sad stupid face.

— I'm sorry. I've been going on. I didn't mean to go on like this. And even you. I know you probably, you probably, had some, feeling for her too. But let me just tell you what I have to say and then this — what did you call it? — cup of tea will be over.

Howard knocks his mug of tea to the floor and watches the spill spread like a scientifically-minded child would. When the lake of tea reaches its limits he looks at the poster again and starts to speak.

— You. You should have been here. I called the factory. I waited for you. I couldn't think in that way. Unanswerable. I didn't know what to do. For once in my life I needed you.

Where were you? Fucking that nurse I expect. I'm not looking for consolation. I've got people I love for that.

Thank you very much, you whining fuckhead. Thank you very much for nothing.

— Look. Howard. How about this? Allow me to suggest a little something. Why don't you get to the fucking point?

— Gerald.

— Who the fuck's Gerald?

— He's a porter or an orderly or something here. He must have spent a lot of time with my, my mother.

Howard, inadvertently, is about to take Charlie's advice. Howard is about to cry. It is a race between the start of the tears and the end of his story. Charlie's money is on the tears.

— I found —

And the tears start. Rich, magnificent. No real change to Howard's face, it just settles into a slightly surprised resignation as the tears rush down. Charlie offers Howard his handkerchief. Howard takes it.

— I'm sorry. I can finish. Thank you. There's been a will. She changed her will. Wrote it out. Left everything to Gerald. That's it.

— That's —

— I found it. I didn't know what to do. I called you. You'd've known what to do. It was in her bedside table. Hand-written. Spelling mistakes. They left her *effects* for me to sort out. Books and chocolates, a few clothes, and tissues and stuff. A jewellery box that was empty . . .

Howard pulls open the carrier bag he has been holding all this time. He allows Charlie a glimpse in at the worthless things that Dinah has left behind.

— . . . And this will. In between a couple of murder mysteries. Waiting to be discovered. And then Gerald, didn't know his name then, probably should've, saw him in passing a lot, silly not to've talked to him before. He came in while I was looking through her things. Supercilious. Offered me a cigarette in this sort of Noël Coward way. *Suave*. He waited until I found it. This will. Left everything to him. I tried to call you. He stuck by my side, grinning, trying to make conversation about the weather, and politics if you believe. He asked me to give it to

319

him. I thought maybe you'd've said burn it or something. I didn't know. He asked for something. I gave it.

The tears have stopped. That pointless dignity endures. Charlie stares at Howard. At this blood-connected man without the guts or the wit to protect his own.

— End of story. That's it. That's all. Over.

— What do you mean — that's all? There's some suave bloke swanning around with something that belongs to us, to *you*, to Simon probably in the end. What are we going to do about it?

Howard says nothing. He looks at his right hand, at the strand of hair between his thumb and forefinger that he has foolishly rubbed away from his head. He lets the hair drift to the floor only recently sanctified by Charlie and Nurse Mary. He clasps both hands around the bag again.

The solution attaches itself to Charlie like a lover's lips. Charlie lays a gentle arm around his cousin's shoulders, ignores the shudder of aversion that he receives, just hugs Howard tighter.

— Howard. Laddie. Inspiration. Let Charlie look after this. That's what you want, isn't it? You go home and do your grieving and let old Charlie deal with things. Here's a guarantee for you. I'll remove Gerald from the picture. I'll do it so quickly and so well that no one'll ever know he was ever in it. And in return. And in return, you let me buy your shares in the company. How about it?

— I thought before you wanted to sell your shares.

— I know you did.

— I thought the table-tennis deal was a joke. You want my shares? I don't care. Get rid of Gerald. You can have them. I want something as well.

— Name it.

— Well it's five things actually. But really one thing. One sort of thing.

Howard giggles. He is going to cry again.

— Sure, Howard. Of course. Whatever. We'll talk about it tomorrow. Whatever you want.

— I think I'm going to go home now. There'll be a funeral, I suppose. Could you leave me alone now please.

A pleasure. Weep now, will you, for ever.

4

CHARLIE, CHEATED CHARLIE, thoughtful Charlie, angry Charlie, strides along hospital corridors. Take stock. Howard still has his twenty-four per cent. Dinah, in a dodgy will, has left her two to Gerald. First things first. Find Gerald. See what he's up to. Sort him out. Buy him out of the picture probably. Then find out what the real will says. What's the best thing that can happen? Gerald pisses off for a chicken shit amount and Dinah left her shares to Charlie. What's the worst? Will goes into probate, nothing can happen until it's resolved, Chatterjee's people go elsewhere, fortune's moment is over. Tempting to buy up Gerald, support the new will, carry on with seventy-six per cent in his pocket. There would be something not right about that. Can't just let a stranger muscle in. Where's Gerald? Staff Nurse Mary McDonnell can't be tracked down, too busy angelling in the wards. A student nurse points him in the direction of Maintenance Engineer Gerald.

That's the lad, lounging with a mop and bucket by the fire exit stairs. He's sitting against the sill of a frosted window, arranging cigarettes. Thinning blond hair. Face of a juvenile lead turned into a character actor by his liking for the booze. Young bulky bloke. Red supercilious face. Looks like a nutter or an actor or a born-again Christian.

— Weather might be about to turn.

Pleasant observation from Charlie. Gerald looks up mildly at Charlie.

— I think you're right. Rain's on its way.

— This sunshine's too good to last.

— Got to make the most of it when you can. Remember the summer? I don't. I blinked and missed it.

— If it wasn't for the rain we wouldn't have rainbows.

Eh? Where did that come from? No matter. That's the weather taken care of. What other pleasantries had the boy bothered Howard with? Politics? Charlie can do politics.

— They say there's going to be a cabinet reshuffle.

The boy sniffs.

— Garbage in, garbage out. They're all the same.

— Some worse than others, says relativist Charlie.

— All as bad as each other, says absolutist Gerald.

— You might be right.

— I know I'm right.

— Got a spare cigarette on you? (Charlie isn't sure he knows how to pronounce the word in a Noël Coward way so he doesn't bother to try.) My name's Charlie.

— No it's not. It's Lawrence.

Gerald smiles smugly. He arranges his packs in front of him, he has about ten different brands.

— Take your pick.

Charlie resists the temptation to punch the boy. He points at the Camel filters, and Gerald offers him the pack. Charlie lifts a cigarette clear and as he is moving it towards his mouth a lighted match appears as if by magic at tip's end. The boy is a conjuror too and that makes Charlie hate him even more.

— Uhm. Thank you. Very clever. Doesn't taste like a Camel.

Nor is it. Some blurred writing on it, might be Arabic. Gerald shows him more cigarettes from his collection, lifts them away from the packs, Silk Cut, Marlboro, Benson & Hedges, they're all identical.

— I get them from a supplier in Queensway. You're a connoisseur. Most people are fooled by the packaging. I'm impressed. *Lawrence.*

He puts a nasty edge on the buried name. Charlie smiles, refuses to show him how much he detests him.

— How'd you know my name? I don't believe we've met.

— Oh we've met all right. I've seen you around here, you just never noticed me. Lawrence Levy. Born 1946. Father Sam, mother Esther. You don't have any kids, do you? Live at (memory man routine, screwing up his eyes to aid

322

concentration) River Apartments, South Chelsea Wharf.
Number thirty-seven?

— Seventy-three actually but very good all the same. How
do you know all that?

The wanker'd got the apartment number right but Charlie
can't allow him to feel too on top of things.

— I was a friend to your aunt in her last few months. She
was a lovely charming lady. Her family was very important to
her. More important, I sometimes got the impression, I hope this
doesn't sound too presumptuous, than it was to her family . . .

Sanctimonious git.

— . . . Howard Levy. Born 16 November 1948. Father
Daniel. Mother, bless her (he crosses himself extravagantly),
Dinah. Two children, Ruth and Simon. Lives somewhere with
an odd name, Chumpstek? Saw him earlier. He's a bit of a
tosser if you don't mind me saying.

— Don't mind at all. I share the opinion.

Actually Charlie does mind, surprisingly. It's all very well for
him to loathe Howard but this sleazy arsehole with his poncey
manners — probably a failed actor, wouldn't be surprised — is
threatening his family.

— I'm sorry. That was rude. I shouldn't have been rude.
Especially at a time like this. I'd like, if no one minded, to
attend the funeral.

— You better give me your address.

— Oh no. I'm not falling for that one.

— I'm not with you. Which one?

— *I'll* get in touch with *you*. I can probably find out anyway.
Jewish place, I expect.

— Expect so. Look. Gerald. Perhaps we should talk
straight . . .? You've got something. A piece of paper. Written
by a derelict old woman going bananas.

— She was a fighter. You shouldn't talk of her in that way.
And actually I think it's called a will. A last will and testament.
The final disposition of a dead person's estate. Usually a matter
for respect.

— It won't stand up. You know that. I know that. Dinah
probably knew that. That's if it exists even. I don't believe I've
actually seen this thing.

323

— Of course.

Gerald reaches into one of his many pockets and pulls out a piece of paper. Holds it high like a trophy. His head even bows slightly in bogus humility. Charlie reaches towards it and Gerald — eyes down, not looking — pulls it back, keeps it the same distance between, close enough to read, not close enough to touch. It's a piece of her writing paper. Familiar address heading in that thin red type she liked. Big words fill the page. Dinah's writing. Dinah's quavery signature at the bottom.

— And it's witnessed.

Gerald flips the paper slickly around. Two signatures there. Signatories' names in capitals. Miles of medical qualifications after them. Charlie is starting to get impatient.

— That's crap too. You know how she was. Doolally. I loved the old lady but she was out to fucking lunch by the end. Go to the law with that thing and they'll laugh at you. Charge you with extortion. The hospital investigates you. You lose your job. Then the papers get hold of it.

The kid thinks he's in a war movie. He stands militarily erect, mop and bucket beside him, cigarettes around him, frosted window behind him, his ticket into Charlie's world held defiantly towards his evil foreign enemy. His words are full of pluck.

— Are you threatening me?

— Just trying to be patient actually. That's the situation. Let's deal with it. Why don't you tell me, Gerald, what it is you want.

Gerald relaxes. Stands at ease. Supercilious smile.

— I don't think it's any of your business. Are you trying to offer me something?

— What do you *want*? Just tell me and I'll sort it. Let's get this over with. I'm tired. I'd quite like to go home.

— I'll be honest with you, *Lawrence*. Money is important to me. No. Hear me out. Of course it's important. It's an unlucky world I've discovered. There's people who have and they keep it all. Start with nothing and they'll try to convince you you're better off with less. It costs to breathe these days. When I was a kid I think things used to be nicer. But that's how it is. Some

324

people get just one opportunity and they either grab it or don't. I suppose what I'm saying is, I'm weighing up my options.

A dashing smile. A coin lifted from behind his ear which dances back and forth across his knuckles.

— When you've decided let me know.

Charlie yawns. Kicks away the butt of the Arabic cigarette which he dropped some time ago and is still burning. The hospital changes at night, becomes a ghost of itself to haunt the dead and the dying. No one else in sight down the length of the corridor. He needs some weed to take the edge off the coke. He needs more coke to get him out of here. Fuck Gerald. He'll deal with him another time. Perhaps Staff Nurse Mary McDonnell is due to finish her shift? Comfort can be found with her. Charlie can't bear the thought of searching. He's been too long in this place already. Is it Wednesday? If it is he's missed his aromatherapy appointment. Nothing else planned. Even Mark doesn't know where he is. He'll have to try and make it home by himself. The car's downstairs, that's too far away. Everything's too far away. Concentrate. There's the nurses' station. The lift bank behind. Basement. Car park. If he can get to his car he'll be safe. All it is is a matter of finding a lift and travelling in it. If a thousand morons can do it every day then Charlie can do it. He wishes the world weren't so quiet at night. His blood is too loud. He needs something else to listen to. Corridor, lift, corridor, basement, car. He can do that.

— All right. Listen. Listen to me!

Gerald there, furious at being ignored.

— Are you listening to me?! I want to tell you something. Just to keep breathing you have to keep chiselling for chances, and when you get one you're a damned fool if you don't take it for everything. I'm a businessman, too, Larry. Not as big a one as you, of course not. I haven't got your natural advantages. But even so I'd risk it all, enjoy success even more. If I can make trouble for people like you.

Where is he? There's the nurses' station. From there it's not far to the lifts. Charlie wishes his feet didn't hurt. That's a new one. Inga has been nagging him to let her loose on his feet, try out her reflexology.

Gerald caches Dinah's will inside a different pocket. From a pocket below Gerald fetches out a multi-coloured pack of Sobranie cocktail cigarettes.

— These are the ones I smoke myself. And (a charming society host's smile) the ones I offer to my friends. Your lawyer — Katz is it? — will be hearing from mine.

5

IT IS HIS silent torture, his secret martyrdom. Perhaps Mark suspects. Probably Inga too but he hasn't even dared tell her. To talk about it would make it too strong. Charlie thinks back to the time before. When he didn't spend every day waiting for fear, his likely companion. Before that moment five years ago when he stumbled into the hospital at night to complain of chest pains. He was sure it was a heart attack. They had checked it all out, with listening devices and electrical currents, told him it was a panic attack. It happens all the time, classic male symptoms, all men are hypochondriacs, trained to experience anxiety as imminent death. But what if — and this might be the fear's question, not his — what if he has developed a heart condition now *as well as* panic attacks? Inga warns him, she keeps saying, *Lay off the drugs, fatty man, the booze, the smokes, make your exercise gradual, not sudden.* There would be an irony: his body finally catching up with his fears. Forget it. If there's one good thing about facing imaginary fears it means that he can feed his body with some real stimulants and fuck the consequences. The edge. Before this, what was he? A slick salesman with two difficult ex-wives. What is he now? Trembling, breathless, aching, and about to be free.

He makes it, Charlie does, across London alone, the windows up, the radio as loud as it can go, driving hard rock at him to remind him to stay in lane and stop when the lights turn red. He doesn't permit himself any time in the garage, he could stay there for ever, timid, hiding, dead. Into the building he goes, past TV-watching security guard French Terry, marble floor, speedy elevator, up, into his apartment and into the bruised diffident arms of Rocky Sporgente's girlfriend, who is somehow, magically, waiting for him.

It takes all of Charlie's ingenuity — lines of coke and heroin in surprising places; the constant threat, or promise, of violence — to make her do just what he wants her to do: to hold him, to hug. But when she finally gets wise to his unsordid game, she pulls away from him, adjusts her clothing, prepares to leave him there, banked on the Chesterfield, alone.

— I'm going to America.

Nothing to say to that. People don't always get to America.

— Rocky's got his championship fight. Atlantic City. He wants me to be there.

— Uh huh. When you going?

— A week or two. But he's fighting here first, a warm-up. He has to win this one. It's for television. I have to be there too.

— Sure. Cheer on the boy. Come here.

— No. I don't like you like this.

— Like what?

— Needy. It's not what you're good at.

6

THE SERVICE IS boring. Charlie is a little stoned and his funeral jacket is tighter than it was the last time he wore it and he can't follow what's going on, neither in the prayer book nor on the Order of Service pamphlet. The pamphlet, Hebrew on one side with a phonetic crib sheet on each facing page, is a shoddy piece of work. The tails of all the *y*s and *j*s are broken. The *e*s and *a*s are blobby little circles filled with ink. Charlie lifts up the staples on the middle page, the pages work free, an act of sabotage that makes him feel something has been achieved. Leslie beside him is in his element, he knows the prayers off by heart, swaying and weeping, Death's vibrato-voiced chorus boy. Howard on the other side, standing tall and alone at the head of the hierarchy of grief, is perfectly still, perfectly silent. Like a lion tamer trying to keep his beast in its cage, Charlie spends the time fighting to bottle up a fart. The men are on one side of the hall, the women on the other. No talent over there. Foundation-faced clown-witches in brown or yellow fright wigs worn atilt, unbelievable smell of perfume failing to cover the decay. Vivien's looking horny though. She's hit the right note of grief and reserve. Tessa next to her, never understood Jewish aesthetics, has draped herself inside miles of foolish black lace. She's getting big, 'll have to watch herself, putting on weight around the middle there, honey — or, she couldn't be, the doctors said it was impossible, fat with baby at last? The rest of the women, most are small, most are fat, most have pendulous lower lips dripping with salivary sorrow and carry paper tissues crumpled inside a hand. Where have all these people come from? Bridge clubs and charity committees, Temple Fortune and St John's Wood and Golders Green and

Edgware, let out for the day from suburban shtetl darkness into bright cemetery light.

— She had, the rabbi informs his captive congregation, beginning to sweat now in the over-crowded heat, a gift for friendship.

Good, fruity singer, over-clipped beard, fine actorly tones in his peroration, he's fond of words which are awash with waves of syllables that he can surf his voice over, verbal adventures, a catch in his voice that he overuses in words like *devotion* and *virtuous* and *righteousness* and *family*. He uses *good* now and again, but only when he's run out of alternatives, and the word comes out with a crispy sound, a pucker of his lips, disappointment, even apology in his eyes. Little mean-mouthed men in top hats, the rabbi's sidekicks, wait by the door, their parts in the act are about to start.

The jacket pinches in the shoulders. Charlie rolls his head like a boxer. It's annoying being stoned like this, half-way house, need to be completely stoned or not at all. In his jacket pocket there is another ready-rolled reefer. Love to sneak around the back of the crowd and smoke it. Where's Ruthie? She'll always join him in a smoke. Next to the reefer is the photograph. Charlie knows it's a luck charm. He just can't decide whether it's for good or bad.

The rabbi smacks his lips together, sucks on an invisible cigarette. The indoor part of the service is over. Now we're going to get al fresco. Charlie releases the tip of his fart as members of the congregation test each other's early judgements on the rabbi's performance. The tip eases out, and the rest follows, noiselessly and no need to worry about the smell, there's enough Chanel to cover a thousand farts. Husbands and wives meet eyes or avoid them, according to the state of their marriages. Widows cluck and dab eyes. Long-lost cousins grab rough hold of one another. Tessa, triumphant, rubs her belly and simpers out a smile. On a wide pallet of wood in the middle of the hall there's a dark wood coffin which, curiously, Charlie hasn't noticed up till now.

The little top hats cluster around the principal mourners, pull at their sleeves — *You. Yes. You there. And you there. There!* — arrange them around the coffin, grab their hands, force them to clasp an end each of one of the poles, Leslie and Howard at the front, left and right, Charlie behind Howard, is that cancerous Leon from Australia on the other side? the weeping elephant Mr Feldman at the rear, and opposite him a large man wearing a dandified waistcoat who Charlie has never seen before. The top hats whisper among themselves, point at watches and, simultaneously, make Mussolini salutes with their right hands, the signal to lift.

They walk slowly, bumpily, from the hall out into the car park, a sheeps' procession between Volvos and Mercedeses, past Mark sleeping in the driver's seat of the Scimitar, the rabbi stepping out their stately path ahead, their progress herded by the top hats yapping around them. The six men carrying the coffin (hard to keep pace with Howard, irregular giant steps, takes three of Charlie's steps to each unpredictable two of the chief mourner) falter, lurch, continue: a quick shuffle through the black iron gateway, check to see if the road is busy, just a postman making his difficult way, pushing his bicycle up the hill, then on, the rabbi leads them through, the top hats scurry around like dogs, behind the coffin come the rest of the grievers, the men and then the women, walking with stern expressions, blinking from the sun.

Sweaty work this, thirsty, would like to stop, can't, the coffin would slide out of the gate, down the hill, break the surprised postman in half: on, into the New Cemetery (consecrated 1946), stop at a sink protected by a little wooden shelter, here we must wash our hands and pretend to be mouthing the appropriate ritual prayer. And now along the cement path that divides the acres of headstones, pebbles scattered on the tombs, grief's useless harvest, a few flowers among them forcing shocking shafts of colour through all the grey and brown and black.

The rabbi leads them down to the bottom of the cemetery meadow where an open grave awaits beside one that's been closed for nearly half a century. The rabbi nods. The top hats practise flagless semaphore. They have reached the place. Sam and Esther are buried somewhere near here, Charlie's mother took him here once to see his father, forced a stone into his unwilling hand, gripped his wrist to make him drop it on the slab; Charlie, Lawrence in a velvet suit, had started to cry and refused ever to return here again. But he has returned, twice, to watch his mother's coffin take its turn to be lowered into the ground, and then thirteen years later for Freda. Now he's back for a final time and he vows to make himself remember to change his own will, put it at the very fucking top: cremation insisted upon otherwise this document's void, ashes to be scattered at sea,

No headstone yet for Dinah, her corpse will have to serve its year's probation first. To keep her company shall be Danny's bag of bones, a couple of pebbles resting on the marble slab that conceals it, his 1951 headstone still easy to read, the dirt of time has been scrupulously held back by Dinah's careful fingernails. (Who will tend it now?) Brown earth has been displaced into mounds along the side, a shovel wedged into the ground between them.

The coffin is slowly lowered, the rabbi approving the mechanics of descent with a proud engineer's expression. Leslie and Mr Feldman weep and compare handkerchiefs. There's the mystery man with the waistcoat, fancy dan, talking to Dinah's chiropodist, a little Indian guy, Chatterjee lookalike, with his clever Asiatic fingers hiding in his trouser pockets. Charlie's seen the transaction details of Dinah's bank account. She visited a reflexologist too, once a week. Like some women have for hair Dinah had for feet.

The rabbi looks around, pretends to be making his special spiritual connection with each individual heart when really he's counting heads. Top hats whisper and compare watches. A couple of slowpokes in wheelchairs are holding up proceedings. At the far side of the graveyard, where the dog noises are coming from, a bare-chested gravedigger in blue jeans worn low lights a cigarette and leans upon his spade. What's Howard doing? Nothing. Pretending to be there. Top hats break out of their huddle.

More singing and chanting to come. Leslie throws himself into it, heart, soul, handkerchief and larynx. Howard stands with his hands clasped in front of his suit. His head is slightly bowed. His eyes are shut like he's praying, probably for it all to be over — the party's going to get intimate soon, and fed, round at Howie's place. Vivien moves to stand next to Howard, to be his rock, and two top hats usher her impatiently back to the purdah of women. Dogs howl near to them. Charlie squints his focus over the bewigged heads of women and the be(paper)yarmulked heads of the men, over the dormitory bedheads of gravestones, to an orange sign fixed to the high wire fence that marks off the next field: *POOCH MOTEL* in big black letters and underneath, in fake medieval script, *dog kennelling service.*

There he is, he did come after all, nasty plump Gerald clutching his yarmulke with his right hand, trying to keep it from slithering off his pink round head. In a brown suit that remembers his body when it was thinner, he's standing discreetly near to the back of the crowd, a few little moustachioed card sharpers around him, Dinah's old gigolos of the baize surrounding the viper, Lenny Leverton's spiritual heir, the pretender to Charlie's fortunes. And behind him, where did they come from? two oriental boys, dressed like waiters, Japanese, Charlie guesses, standing shoulder to shoulder, heads slightly lowered, bodies poised and dancer-still, looking fierce and graceful.

Howard beside him has his sleeves tugged by two top hats, they pull him towards the shovel, ratty little men make rodenty gestures to mime what he is to do. Howard takes the shovel, pushes it into the nearest mound of earth and shakes the first few sods on top of his mother's coffin. Howard frowns. The top hats try to rescue the shovel from him, he shrugs them off, they pull at his arms, he ignores them, sends a bigger load this time, his covering gift to his mother, consents finally to let go of the shovel. Leslie's turn now. He steps chirpily forward. Charlie will be next.

Howard dusts down his trouser legs. He looks up, hopeful, sniffing. As if he expects to detect something miraculous. But there is no change in the weather. No angels singing, no warm sudden breeze or flash of lightning in the sky. The clouds do not pull apart to reveal something fuzzily divine. Charlie is offered the shovel. Fuck etiquette, he's bigger than he used to be, this jacket pinches. He can't risk it ripping in the exertion. He shakes out of it, passes it to Howard to hold.

It leaves behind little flakes of itself, in jacket pockets, in attaché case corners. Its dead photographic children which curl into dust. Sam's legs are almost gone. Danny's hair is daggered with white. We need a restorer. A man with expert fingers who will sit at his well-lit work table and cheat with inks and spray. Perhaps he will dismount the fragments, Leslie's inadequate work, soothe them back together and take a photograph of that.

But then it may become less than itself. The magic of this photograph derives its power, at least in part, from its material history. From its earliest, glossily wet moments, pinned up to dry in the back room of the portraitist's shop at Manor House. Then, in a broad mahogany frame, it sat in the parlour of Dinah and Danny's house in Stamford Hill. Before he died it had already fallen out of favour, it went to Esther's house, to LPC (to hang in the boardroom beside an ageing moulded button), then back to Dinah and then away, hidden inside a storage box. It has been touched by the men themselves: by the nervous clever hands of Daniel, by the strong, slightly brutish fingers of Sam. And by their descendants, Howard and Charlie, Simon. Its magic does not work on anyone outside the family. See now? Howard drops Charlie's jacket because he has forgotten, in his grief, that he was carrying it. Someone picks it up for him, this someone's fingers might even touch the photograph (or at least the ruined remnants of it — when does it become so badly disfigured that it is no longer itself?), and there is nothing. But give the jacket back to Howard again. Have his fingers accidentally dip inside, to fidget with a grass reefer, have his fingers, as if by magic, discover the photograph, have him, wonderingly, withdraw it (and the jacket hits the ground again, we have no need for jackets), touch it, examine it, and for a while at least we are with him again, the family line continues.

EIGHT

SUCCESS

The route to success is seldom obvious, even to hindsight's genial gaze. We think of short-sighted Maxie Leibnez in his laboratory, the invention of PPC just one clumsy stumble away. We think of the first washing-up bowls, the housewife's sturdy friend, and the gay hula hoops that all little girls love, and we remember with a chuckle that both came to being because of the surplus of Polythene in U.S. quartermasters' stores after the Second World War had come to its grim end.

H. C. Willi, *Century of Plastic: A Celebration*

I

FROM THE FUNERAL they drive in procession back to the house. Howard and Vivien sit in the back of the leading car. The gates are open and Ruthie is waiting for them in the drive. Dressed for suffering in a long black dress, she holds a posy of white flowers torn from the garden.

More cars come sadly up the drive. Ruthie whispers to her parents as they walk towards the house on either side of her.

— I wish I could have been there. Was it horrible?

The mourners step out of the cars in tactful, jacket-buttoning slow motion. There's the rabbi and there's Charlie, red-faced, panting, talking to pregnant Tessa, and Bertha there and the chiropodist there and Leslie there, dabbing at the moist parts of his face, and is this what he returned for?

— It's the way it is, says Vivien. If your parents are alive you're not allowed to the cemetery.

A corpse is an unwieldy entrance ticket. Do they ever check the funeral-goers' credentials? *Hey Solly! Over here! I think we got a gate-crasher! ... C'mon, you. I'm talking to you. Show us your dead parent or get the fuck out of here! You don't belong ... That's right, Solly. You tell him. Tell it to him good ...* Jewish irony to choose that as the barricade to erect between childhood and maturity. Today I am become a man. Today I enter the cemetery of adulthood.

— It was horrible, says Howard.

His daughter reaches for his hand and finds it and together they walk into the house.

He escapes, if only to the bedroom where his wife discovers him lying spreadeagled and shoeless on the bed.

— Are you all right? You'd better come down, they're doing some of their mumbo jumbo stuff. I think you're meant to be there.

— I'm coming, he says and stays where he is. I thought Simon would come to the funeral. I don't know why.

— So did I, she says.

She sits on the edge of the bed and primps her hair in the dressing table mirror. He watches the things her shoulder blades do through the thin black silk of her funeral dress.

— They used to be very fond of each other.

She turns her head to say this to him but the words come out blurred through the three hairclips she holds between her clenched teeth.

— That stuff only really works, he says, the mumbo jumbo stuff, if you believe in it. Otherwise what's the point of it?

— Aren't you meant to be crying in public or something? Showing your sorrow to old women with moustaches who knew your mother? I thought it was meant to be cathartic.

He would like to respond to that. He would like to say, *Well, I'm crying in private, isn't that good enough?* but that's far too glib and far too sentimental and anyway he's suddenly crying too hard to say anything. It is wonderful, unstoppable, after a brief pointless attempt at control he just gives himself up to it, and the tears keep coming, voluptuous.

Vivien is beside him, kneeling, her hands behind her head as she fixes the last of the pins into her hair.

— Oh Howard.

He is panting, letting the noises gush out from their gut-soul place, groanings and weepings and gasps. Maybe this is what he was looking for when he went searching through his selves. It's not even sorrow, it's closer to liberation or madness. His fists clench with every shudder of his chest and belly. Fingernails tear his skin.

— You've made a cut.

He opens and closes his eyes. The image of his wife's face is gone, here, gone. Magical power that he has.

— Here. Let me.

She's pulling at the fingers of one of his hands, he's not even sure which one it is. She's licking a wound and looking at him.

Her tongue laps against his skin. He grabs at her with his free hand.

— My dress, she says, Oh.

His trousers are down, her dress is billowing up. They make love on their marriage bed inside a canopy of funeral silk. Somewhere a rabbi intones, chants.

— I, he says, Need, he says, Please, he says.

He is not trying to tell her anything. They are just words, like breaths. He forgets to cry. *This*, his body tells him, *this! is what we're for* . . .

They are quiet. Carelessly, she waves towards the door.

— Shouldn't we . . .? Isn't it . . .?

— Yes.

Their clothes are ravelled up together. They lie, she on top, thigh to thigh. Her left hand is caught between them.

— It's all going to . . . I have to go to the bathroom. Or something.

He wraps his arms around her. Familiar unfamiliar flesh. This ripple red-marked from waistband elastic. The raised birthmark on the western edge of her coccyx. The tiny ridge of scar from where her perineum tore when Ruthie was fighting to be born.

— Please. I tell you. It'll make a mess of our clothes. Everyone will see.

He continues to hold her tight. This spot just below her ribcage which has always been for her unbearably ticklish. His heart could lurch now into either sadness or glory. She yawns and so does he. She pulls away her trapped hand and moves both hands to cover her mouth and she rocks from side to side upon him, a reassuring wetness spreads, his come and hers mixed together.

— Before, he says. When I went past, you know, his door? His tray was still there. He hadn't touched it. I was worried.

He'd thought it was Simon's way of joining in the mourning. *Can't come out of my room, couldn't do that, but I'll go without dinner for one night, how I loved my Granny, isn't this the grandest way to demonstrate my devotion?*

She makes to wriggle away. He prevents her.

— I thought he might be sick even.

— He's not sick. He's done that before. More and more the last few months. Haven't you noticed...? He's all right. Howard. Let me go. We should...

She pushes harder against him. This time he lets her go. She pushes herself away and walks on her knees to the edge of the bed. He lies with his hands clasped behind his head. His elbows point forwards. He waggles his arms like failed wings. When she returns from the bathroom he watches her reflection in the dressing table mirror. She smoothes her dress down over her legs. She retouches her face with powder.

— How do you feel?

Her mouth moves exaggeratedly, as if she's talking to a deaf person who lip reads.

— I feel fine, he says.

— You should get yourself together. We should go downstairs.

He hitches his trousers up. There is a smile on his face. It feels right to do this. To be with his wife, to fuck his wife, to be asked by her how he feels. It's taken his mother's death for things to find their place.

— Ruthie was looking good, he says.

— You must be nicer to her. She feels out of things.

— She was looking very good.

— We should go downstairs.

— Everything's all right now. Isn't it?

Instead of answering she discovers a smudge of lipstick which has strayed onto her cheek. She screws up the corner of a tissue and wettens it with her mouth and dabs her cheek clean.

— Isn't it?

She shakes her head.

— Your flies are undone. I'll do whatever.

She shrugs, sighs, he had never realised how active her shoulder blades are. She puts her hands down on the bed with straight arms behind her to support her weight. If he wanted he could flick with either foot and kick away a supporting arm and watch her fall.

— You know that. Don't you? Whatever you need. I'll help. But I don't want us to stay together.

— What? *What?*

— Not any longer. It's not what I want.

— So what about that?

He opens his hand, holds his palm in the direction of the past, where lovemaking takes place.

— Was that just because you felt sorry for me?

— It had nothing to do with that. Actually I don't even feel sorry for you. Sorry for myself maybe. Sorry for everybody else. Not you.

She gives her hair a final pat. He kicks forward with his right leg and she must have seen it coming, a movement in his mirror eyes perhaps, because she eases away from the bed a fraction of a second before he would have collapsed her supporting arm.

— Who, where, what happens now? *Miles Rodgers?*

— Probably not. I don't know. He's rather dull. We listen to Broadway musicals together and it feels nice though. He knows a lot about musicals.

— And that's the reason? Come on.

— It's not that I don't love you. It's not. I just wish you had never come back. I'm sorry. I am.

He hopes she won't say anything more. She has more to say.

— It's shown me. You need someone subservient utterly or else permanent make-believe. I won't leave you in the lurch. You can count on me for whatever. I just preferred it when I was writing letters to you. I'm sorry.

For something to do, he replaces his shoes. She drapes the dressing table mirror with a sheet that once was used for Simon's bed, and before that, before children, before tiny scars in perineums and loss of hair, it was the sheet that covered the guest bed in their first married home. Howard ties his shoes and moves to the door. They walk out of the bedroom.

— I'll catch you up.

He watches her take the stairs. He waits until she is out of sight and he waits until her footsteps have been eaten by the hubbub of mourners below and then he goes to sit at Simon's door.

— When you were born, whispers Howard, you had, like, a, pinhead.

There is a tray from lunch, prepared in advance for Simon to eat while they were at the cemetery, a pair of tuna fish sandwiches (fresh) wrapped in clingfilm, an apple and a Mars bar beside. It is untouched, a mark of respect perhaps — this could be his gesture of sorrow. Or he might be sick, gasping silently for breath and dying. Or maybe, like his father, he doesn't eat tuna fish sandwiches.

— It was the labour, you had to be turned around, breech or something, opposite posterior? You had to be manipulated, you were facing the wrong way. Maybe you just didn't want to come out. That would make sense, wouldn't it? Your mother screaming. They tried to kick me out of the delivery room, I wouldn't have it. You were blue at the edges and your head was, sorry to laugh, ridiculous.

My son, Howard had said to himself at the time, sure that this was the bad thing his whole life had been a silent preparation for, *is deformed, horribly.* He had reached for him none the less, blue-pink pinhead boy in a yellow freakshow blanket, duelled for him with a nurse who had lunged him against her own immense bosom, while Vivien tried to push through the drugs and make sense of things. Howard had won the battle, clutched his son, stroked his blue fingers, his son was deformed and hideous and wriggling but so what. Howard had closed his eyes, hoping that would do something to quieten the desperate tenderness that was scarring his heart.

— *My* father was dead before I was more than a year old. Endless stories but I don't remember his touch. I thought that meant I was free in some way, and obliged, to become something better. What else is there? What can you do? *He* could make something that had never been there before. I can't. I at least should be able to look after the things he left behind. Maybe improve them even. Isn't that it? Isn't that what we have to do? Make the old machinery work? Make it new. Or else just ignore the whole thing. Danny killed himself so he saved me the job. I should ignore his memory, make something new for myself. I'm tired. I thought you were going to come out today. I was sure of it. She would have wanted you to be there. Dinah.

Not quite silence. Is that the sound of the boy shifting around in the room? Or it could be a bird, its wings fluttering against the window outside.

— I expected to see you. During the service or after. I shovelled dirt on top of her coffin because that's what I was told to

342

do. Weird, I shovelled on some, it wasn't enough, my mother deserved more, that wasn't anybody, it was Dinah, she was special. So I went for a second go, did better this time, got a ton of earth on my shovel, a few stones sticking out, threw them over, yes, felt good, rattled the load on top of the, her, coffin — and these repulsive little men in top hats, they were running the show like little munchkins, they tried to stop me because I was taking too long or maybe it was because I'd exceeded my ration of dirt. I wanted to hit them. I've never had such violent thoughts before . . .

Howard acts on a foolish inspiration and checks the door knob. It turns . . . it might be . . .? Of course not. It is never that easy.

— You're right, I probably have, Miles Rodgers, wanted to hit him, and Charlie . . . Your mother doesn't want me any more. I have to go downstairs now, had to tell you this — after the palaver with the shovel, I went back to my position, stood there, closed my eyes, and waited for you to show. It wasn't that — it's like those old men muttering and jerking their heads and shaking and fiddling with the tassels on their prayer shawls, tzitzits and tallusses and chuchkas, and if they do it enough the messiah will come — I *knew* you were going to be there, I just had to wait. You didn't show. Do dead people get mixed up with the soil? I suppose they must.

Suddenly Howard is almost there. Almost at the truth of what Dinah had been trying to tell him. He can see the shape of it. He can smell the necessity of it. If he could push aside all this rubble in his mind he would be able to apprehend it. Is that a cough? The clearing of a throat in preparation for speech? Howard waits. It seems like he's always waiting. Howard is waiting for his son to speak. Howard has decided his son's voice will be low, mellifluous, calm. He can smell him, his son. The scent of Simon curling out into the world through the crack in his door. Simon, sour beer and liquorice. Howard gets up, taps out a tattoo on the door. It was the rhythm they had agreed on years ago, when Simon was in his room because he was sleeping, and Howard was on his way to work and this was their code. There is no inviting tap back. Howard tests the door with his foot. Quite flimsy really around the lock. If he just pushed, hard, sudden, like this, the lock would splinter, the door swing open on its hinges, to reveal . . .

Unbelievable he hasn't tried it before. Vivien did, she'd written to him about it, it was soon after he'd gone away, they hired a builder to remove the door — and the sight of the top of Simon's head as he hid behind the bed had been enough to break Vivien's heart all over again. She told the builder to rehinge it and lock it all up again, she gave him a tip and they left her son as he wanted to be, protected, alone.

Howard coughs and simultaneously he kicks. The sole of his shoe against where the lock should be. The action doesn't make as much noise as he expected. Nor does it open Simon's door. He kicks again — and he coughs again to cover the sound — and it is so satisfying to feel metal tear away from wood, screws spinning, the lock comes free from the door post, the door swings open, a gust of cold air, the bedroom window is open wide. Howard steps forward, pauses, he dares finally to trespass into his son's hermitage. The room is a mess, books scattered everywhere, sweet wrappers, coffee mugs, clothes in piles like dead men's things, bowls and spoons piled in columns in the sink, an ugly smell rising up from the chemical toilet, scraps of paper with elegant writing upon them all over the floor. Howard walks through the room, the chest of drawers, the book shelves, he closes the window to abate the storm of sweet wrappings and papers. There is a transistor radio on the unmade bed, and no Simon. Howard switches on the radio. A phone-in programme rises weakly up. His son is not in the room.

Charlie is in the hallway. He's acting lewd with an old dear called Muriel. For a moment it looks as if Muriel's about to hit him. Her eyes blink furiously behind thick-lensed glasses. Her mouth, which doesn't quite fill the lipstick heart outlined in brown and painted in carmine, hangs open. Her shoulders flinch. Her moustache bristles. And then Muriel clutches her brown wig with little liver-spotted hands and fills the upper air with merry cackling laughter. Charlie whispers something further and Muriel, one of Dinah's bridge circle pals, cackles and snorts so hard that Howard fears she will break.

People grab at Howard. They want to gush sympathy and

344

mother-praise up at him. He is bent, rubbed, squeezed, pressed, shaken. They touch, keep touching, he doesn't know what for, to take or to give. Vivien has been waiting for him. Howard takes hold of her offered hand and they push through the grievers who make harsh farmyard noises and examine the ornaments.

— The rabbi's gone, she whispers, he couldn't wait any longer. I think you have to sit down now and everyone comes to give you sympathy.

They find their seats. Stools to sit upon on opposite sides of the drawing room. Stools to signify that they are all, equally, lowered by grief. He hasn't told her his discovery. She didn't give him the chance. (Simon's door is shut again, anyone passing will presume it locked.) Howard will be touched even more now, with hands and lips and hair and other people's tears. Everything feels disarranged. His son should be sitting beside him. Charlie's laugh surrounds everything.

In an over-large house in a pretty suburb of London, Howard Levy waits for his son. (Is that him? there, in the new black suit? Or there, with Katz, young accountant of the month?) He watches his mother's mourners. At another time he would have been drunk and flirting. Or soberly examining the future for hopeful signs. Or crying or raging or cursing. Howard lights a cigarette. Howard still smokes cigarettes. He smokes them because his wife, his daughter, the government, tell him he shouldn't. Howard, the onion, has watched his selves peel away like skins. Off they went, discarded, often bringing tears to his eyes.

He wishes someone would tell him what to feel. He wishes he could tell himself. Howard has been hungry for feelings that are indisputably his own. Today though he has found something, two things. His secrets. They give him a pulse of pleasure, almost to joy. He needs to share them. He needs to keep them to himself. (Who can he trust? Nobody. He has learned that much. Maybe Ruthie, and even she would be a risk.) How long has Simon been gone? There was only one untouched meal outside his door.

— A chair for Mr Feldman!

Wisely, Charlie has fetched a stiff-backed chair from the

345

kitchen for the corpulent bookkeeper to rest upon. But Feldman sniffs at it, shakes his head, he wants something more comfortable and damn the escape-proof consequences.

Maybe he went years ago, but his accomplice (Vivien? Ruthie? Charlie's driver?) has dutifully emptied the tray every mealtime. Maybe the myth of the Boy In The Bedroom has been only a myth. Maybe it was a suitable legend to cover up what? Disappearance? Murder? Dinah's death has upset everything. The facade crumbles.

Leslie simpers up, clasps Howard's right hand between both of his. Fixes him with a look of deep sickly sympathy.

— I, says Leslie, my mother also, yes, deeply tragic, for the best. Always, your friend. Whenever. You know that.

A look of utter malice slips across Leslie's grieving face.

— Yes Leslie, I know that.
— Always.
— Yes Leslie.
— Good, sometimes, to weep.
— Thank you, Leslie. I will weep.

Leslie (his own eyes red-rimmed, rheumy, raw, wet) hopefully inspects Howard's eyes (cold, clear, dry) for signs of dampness.

— The Aztecs, confides Leslie, dabbing his face with somebody else's monogrammed handkerchief, God rest her, were terrified of the tomb. Better sometimes I think to cremate.

And off he goes, spins around and spins away, finds Miles Rodgers, a convenient companion to instruct with tales of ancient times.

Howard, the incompetent tear-master of this congregation, is meant to be mourning, open hearted. No vanity or modesty to impede gut-churning high-pitched God-wailing lamentations. (Is God so deaf that He only hears sorrow if it is wailed, lavishly?) The mirrors are covered. The spectators are gathered. Unshaven cheeks that press, bristly, chafing, against mine, wet eyes, bodies rocking within an obstacle course of unwanted, accidental history. There's a business associate, there's a chiropodist. Where's my son? Even if Howard wanted to cry he could not. His tears, finite in number like heartbeats or coins, have all been used up, spent.

— Lovely service. Did Dinah proud.

Charlie up above, leering down.

— Gerald? did you deal with him?

Charlie looks at his watch.

— Gerald. Not yet. In hand.

Charlie's not stoned tonight. No cocaine energy or goofy cannabis charm. It's just Charlie alone, looking old and careful. And he's scrutinising Howard in an unfamiliar way. Usually Charlie's looking at some secret place, in the future, in his imagination, and his patter is all about persuading whomever he's talking at to believe in and look at it too. Close your eyes, children, let's dream we can fly and when we open them again there'll be wings upon our backs, trembling, inexpert, still damp from their ghostly making. So you won't need that wallet will you? It'll only hold you down to the earth, remind you of gravity. But at this moment Charlie is examining Howard as if he is a scientific curiosity.

Howard waves his right hand, princely fashion, airy, towards the bewigged women and the bewhiskered men standing behind Charlie, patiently waiting to rub their rough faces up against Howard's and gum their wet lips into his ears to remind him that time is the great healer and his mother was a marvellous woman, impossible to replace or forget, their dearest friend.

— I have sold the company.

Kiss-whispered by Charlie. *I have*. A gentle kiss, *sold the*, blown by Charlie's thick lips, followed by a sigh, *company*. His face is a mess. Red veins spider over puffy cheeks that are soon to sag into jowls. Charlie stands away a little, smiles a little, inspects, the good social scientist, the effect of his words. He is disappointed. Charlie had been expecting more.

Howard just smiles at him. Howard has secrets.

Behind, the mourners are growing impatient. A damp procession that grumbles out into the hall. Charlie spreads his arms as if it's finally time to fly.

— Do you remember, says Howard, herring days?

It had been their phrase, a connecting pun — Yom Kippur, kipper, herring day. Solemn fatherless days, the annual atoning fast, when the widowed sisters Dinah and Esther kept stern company with their damaged sons. No synagogue or rabbi could get in the way of the absoluteness of the day. It would begin the eve before, Dinah looking at old snapshots of her dead man, Howard sent to bed to contemplate his sins. In the morning, breakfastless and breathless, he'd recite a sleepy and hungry list of every crime he'd committed in the past year while his mother showed no sign of listening. Sometimes, like a Catholic, he'd have to invent a sin in order to repent of it. And later, the only day in the entire year when the prospect of enemy cousin Lawrence was a matter for celebration. It meant that the day was nearly over. A temporary truce, the cousins pleasantly together, released from the terror of their mourning mothers pretending to be God or men.

— Yes, says Charlie. I remember.

And then one year the dreadful day hit silently, not even a whisper to warn of it. Howard gathered his real and imagined sins together, arranged them in a line like toy soldiers and waited for volunteers to announce themselves. And his mother was silent, and the morning toast was waiting, and she was reading her newspaper, and Howard's sins were choking his throat and she watched him eat food and did not ask him to confess. The day passed by, terrifyingly slowly, herring day unannounced, unobserved, and Howard longed for it.

He is free of that now and free of Dinah and there is nothing to fill up the space. Briefly, in freedom, he had learned how to become his possibilities but not here, he can't do it here. Impossible to sustain. The alternative Howards had been murdered too long ago, smothered in their cradles, electrocuted in youth clubs, bled to death in school, in the boardroom, the marriage bed, assassinated on dance floors in party rooms. Their brief resurrections, personality ghosts raised by sex with strangers, had been joyous, uncomfortable, impossible. Here his hair is falling out again.

Charlie. Lawrence. Herring day Harry. Licks his lips. Howard would almost guess he is nervous.

— We're selling the company to Harry Wiener. He wants the trade names and the order book and most of the machines. He knew our fathers.

This life will go on for ever. Until Feldman starts to cry because he's stuck in his chair. Until the mob of mourners behind tramples Charlie down to suck at Howard's soul. Until Howard finds his son again.

— And the site?

Charlie nods. Charlie is like a game show host.

— Good question, says Charlie the Patronising. Charlie smiles. Charlie smirks. Charlie has teeth like a shark's. Charlie has alcohol skin covered with fine hairs like razor blades.

— It's worthless. We always knew that. Too many pollutants in the ground. I've had surveys done. Could never build anything there, not for a hundred years. I'm going to try to get rid of it. I've got a fellow on the Council. It won't be easy.

Charlie examines Howard. He looks surprised, a little disappointed maybe by the slow smile that is opening up on Howard's face. Howard, for perhaps the first time since he returned, since he stood at his unfamiliar front door and waited for his family to reveal itself and let him in again, is happy. A miracle has taken place. His mother is dead. His wife is about to leave him. His daughter is unknowable. His son is gone. He has lost his job. And despite all this is Howard happy? More than happy. Exhilarated, ecstatic. Howard has been released. Howard's father is dead. The past is gone. (If necessary we will deal with the past later. Find an unemployed poet to write the company history. Compose our own recollections of days of discovery, days of hope.) Daniel burns for a second time. For the son to thrive the father must die. Howard is released from Daniel at last. And yet.

— Where's the profit in it for you, Charlie? There must be something?

Charlie just smiles that shark grin again. The crowd presses nearer. They used to belong to the same private tennis club. Howard had the more elegant game, the more powerful serve. Charlie never dropped a set. Time does not exist for Charlie. Only the machinery of power, calibrated, at the tennis club, in points and the quantity of girls' telephone numbers accumulated in his pocket address book. Howard takes five deep breaths. He wipes his forehead with his sleeve. He removes the photograph from his pocket.

— Charlie. I'll tell you what. You can do what you like. Take this photo too. You do want it, don't you?

He holds it out to him; Charlie reaches for it and Howard pulls it away.

— I want only one thing. My father's machines. The ones in Museum Corner. I want them. Yes?

There's Charlie's driver pushing his way through the crowd. There's Vivien hugging Ruthie. There's Feldman sitting down, complacent in the knowledge that the chair is his for the evening. (What happens when he needs to piss? Is there room in all those folds of cloth for a mechanism to catch the urine?) There's Leslie with sympathetic Harry Wiener. There's Bertha sobbing alone. There's Charlie's brother Maurice in his Asian embroidered waistcoat trying to think of adventures to amaze Ruthie with, and maybe waiting to be recognised by his brother. There's a pair of Japanese boys in waiters' outfits. Who brought them? There's Charlie's driver whispering in Charlie's ear. Charlie nods. *Yes. Yes. Yes.* For Charlie everything is yes from now on. *Yes*, he says to his driver. *Yes*, he says to Muriel who knocks against his arm and tells him he is in her way.

— Yes, he says to Howard as he reaches for the photograph.

— Thank you Charlie, says Howard as he lets the photograph go.

2

THE PHOTOGRAPH SNUGGLES *up inside Charlie's breast pocket. If it wasn't for a torn jacket lining, a silk dress shirt, layers of skin reddened and bruised from Charlie's concerned hand, some anaesthetised nerve endings and narrowed blood vessels dammed with insulating dollops of fat, the photograph would be touching Charlie's loudly beating heart.*

Success. To Chatterjee the land. (Signing ceremony to be performed later at the predictable location.) To Howard some rusting machinery. He'll have to clear that with Leslie and somehow connect it to Leslie's necessary punishment, his final challenge before the new life begins.

It will be a relief to get out of this place. Mark steers him through the crowd. Charlie smiles at everyone he passes and mouths the polite words *Fuck You*. Fuck you, Muriel. Fuck you, graceful Vivien. A-a-a-nd fuck you, sweet ignored Ruthie. It's been an avuncular pleasure, really. (Get me out of here, Mark, before I lose control.) And fuck you sideways, Gerald. We'll be getting you later. The basement. As arranged. Wipe that smile off your face. And fuck you too, Leslie, I really don't have the time to stop and chew the fat. Your time will come.

— Charlie, please. A word.

Let me through please I'm stifling.

— I don't. Excuse me. I must —

— A minute only. A word. I have been talking to —

Leslie looks around, like he's in a bad spy thriller, *excuse me, we must be discreet, whisper, our enemy has many wiles . . .*

— Yes? Come on. I don't have time.

— To, to, Our Friend.

— What are you talking about Leslie?

351

— Our Friend.

Is that a nervous tremor around his left eye or is Leslie in fact attempting a wink?

— Friend. *Harry Wiener.*

It's a hard name to hiss out in a whisper and Leslie doesn't manage it, instead attracts the attention of Harry Wiener himself. Wiener looks over at them, the baggy folds of skin around his eyes separate a little to manufacture a quizzical expression. Charlie sends him one of his It's-All-Right-buddy looks back.

— Leslie. Walk with me. Keep me company on the way out.

— No. I must stay. But I have been talking to —

— Harry Wiener. Yes I know. Our Fucking Friend. What's the urgency?

— It is about the machinery, the question of how to dispose of some of the machinery. Our Friend does not want — and can we blame him? sentimental attachment only, I expect — Museum Corner.

My. Sometimes fortune really does favour. Everything fits. This is why he changed his name to Charlie. This sort of thing never happens to men called Lawrence. Lawrence did not ever fit. Lawrences are losers. If you look for glamour, Lawrences at best are sickly men who write dirty novels or dubious types who fancy Arabs and die in motorcycle crashes. Charlies are tree-shakers. Charlie Parker. Charlie Manson. Charles Atlas. Charlie Levy.

— Don't worry about it. I've taken care of it.

— Taken care? How so?

If Charlie was performing with his customary finesse, he would probably dissemble. If Charlie was comporting himself with his usual panache, he would keep Howard out of the picture. Fuck panache, it's only Leslie. And Charlie has to get out of here. His heart is beating too fast.

— Howard gets them. They'll be his cut. I've kept his hands off the rest of the money. Working for our best interests, yours and mine. Look. Have to run. See you anon.

— I don't think so.

— You what?

— Please. Don't misunderstand. Nothing against you, Charlie. Probity, yes. Understood. Integrity. I am afraid though,

352

as a matter of fact, I cannot agree to Howard getting the machines. Not right.

— No money. Just useless machines. Come on.

— I am sorry. No I cannot go along with you if Howard gets machinery. He does not deserve them.

— But they're useless. Junk.

— No matter. No.

Wiener again looks over their way. Charlie wraps his arm around Leslie, gives him a little painful tug on the ear, shows Wiener the thumbs-up. We'll sort this later. No time for it now, and double trouble to come for Leslie, but the Chatterjee deal has to go ahead, nothing can be allowed to jeopardise that.

— Fair enough. I can see your mind's made up. Whatever. I have to go.

Someone else wants to talk to him, Charlie doesn't know who, he pushes him aside and Mark steers him through to the car, a journey that keeps being threatened by people who reach for Charlie with hands and looks and lips. So Yeah, and how are *you* doing matey, and fuck you Bloke with the fancy waistcoat, you look familiar but I just don't have the time. And yes even in the drive outside, fuck you fat Tessa and fuck you everybody and couldn't we go see Inga now? Her clever Swedish fingers?

— I want to go to Scandinavia. Drive me to Scandinavia.

— We have a date in your basement.

— We can keep him waiting.

— And I've got to do something after that. Meet somebody. Come on. Here we are. Are you all right Charlie? Take the door handle. Compose.

— What do you mean — meet somebody?

Panic. Charlie struggles to operate the door pull. He forces his breathing down. It costs him but he slows it down. Mark opens the door for him and Charlie slides gratefully into his seat. He examines his most recent worry. It is dangerous how much he has come to rely on Mark to be his bridge across zones of panic. He'll have to get rid of him. A sign of the new age.

— An appointment. A rendezvous. An assignation. A fucking meeting, understand? You can have the car. Have you got keys?

I better leave you mine. I'll take you anywhere first but after that I'm on my own. All right?

Of course it's not all right. Charlie settles into sullen silence as Mark zips the car out of Howard's drive. Mark switches the radio on and Charlie immediately changes the station. Mark lowers the window and Charlie leans across to raise it.

It is good though in the basement. Almost good enough to forget to anticipate what will come next.

— Hit him again.

Mark hits him again. Open-handed slap to the side of Gerald's face that stings, leaves a red silhouette behind.

— Again.

— Not yet. Wait. I want him to stand up straight first. You ready?

Another red silhouette, pain twins. Gerald is about to cry. Man, this has got to be one of the most pleasurable acts in all existence. Sadism in safety. Enjoy it with your friends. Gerald had been so cocky at the beginning, the ironic English gentleman confident in his power.

— I'm going to hurt your bollocks now.

Mark is good at this, he has a gift for it. Charlie didn't realise he'd be quite this good. He must have hurt people before. He talks more than he usually does, calm though, steady, like a bored tour guide, informative but with no enthusiasm.

— I'm going to squeeze them hard. Like this. And then let go. You've got big bollocks haven't you?

Gerald is beginning to sob. Girlish crying, big gulps for air. Mark lets go, allows Gerald to lean back against the storeroom wall.

— I'm (sob) going to get you (sob) for this.

— I don't think so.

Said by Mark in a considered way. He's calculated the probabilities, he's factored in the variables and, he has to say, that in his expert judgement Gerald is very unlikely ever to get them for this.

— You're going to take your trousers down now.

Gerald shakes his head. Mark nods his. Gerald shakes his head again, more vehemently. Mark inserts his index fingers

354

into both corners of Gerald's mouth and then slowly pulls the mouth into the most painful smile. The sound of skin ripping is not a pleasant sound. Gerald reaches for his trousers. Mark holds the smile where it is. Gerald starts to unfasten his flies. Mark lets go. Gerald drops his trousers. His white boxer shorts are decorated with red polka dots. His legs are pale, almost hairless, marked by varicose veins.

They are in the basement store room of Charlie's building. No one will interrupt them. Charlie has borrowed the key from French Terry the doorman (who is, in fact, Canadian). It is dark in here, smells of damp and rodents. Gas and electricity meters and air conditioning controls line the walls along with wide grey pipes, store cupboards, neglected paintings in deflated bubble-wrap. Gerald, in his own mind what? a superhero? a cool dude? a race warrior? Marlon Brando? had been waiting for them in the vestibule. Charlie already had the key. Gerald wanted negotiations. Gerald wanted money. Charlie offered him a safe place to talk and then he gave him pain.

— Bend over.

Charlie is close to admitting it. This procedure is making him feel sick. He is about to intervene but Mark senses that, looks quickly over, shakes his head, and for a moment Charlie is genuinely confused. He hopes Mark is indicating that everything is going according to plan. He hopes it doesn't mean that Mark is enjoying this and will turn on him next.

Gerald bends over, a wobbling wave of buttock flesh beneath his boxer shorts. Trousers in a puddle around his ankles. The hem of his T-shirt peeps out from below his tailless shirt.

Mark takes out several packs of cigarettes from Gerald's jacket pocket. Mark never smokes, he disapproves. Mark disapproves of anything that shows weakness to the world. Mark disapproves of cigarettes and drugs, and he disapproves of alcohol and guns. Mark disapproves of sufferers of asthma and eczema and he disapproves of people who laugh or smile too much. But Mark lights a cigarette. He lights a cigarette and puffs on it without inhaling.

— I'm going to pull down your knickers now.

Charlie wants to shout, *Enough! We have got what we wanted. He is humiliated, less than human. Stop this!* but

Charlie does not, frightened, partly, of what Mark would do and fascinated, largely, to watch him.

— And then I'm going to brand you on your fat arse.

Cigarette in the middle of his mouth, puffing Mark approaches puffy Gerald. He pulls down the boxer shorts with a flourish of his left hand. Gerald whimpers. He could fight now. He could run. He does not. He remains there, his great moon arse facing the world and his tormentors.

Mark stabs him with the lighted end of the cigarette, first the right cheek, then the left, seems as if he is considering stubbing it out in the anus ashtray in between. Skin sizzles. A moan low in the throat from Gerald who, impressively, hardly flinches.

— Do your trousers up.

Gerald pulls himself up and his trousers too. Nods. His eyes are wet. His face is demonic in the flickering redness of the underground storeroom.

— You see, says Mark to Charlie, it's all about timing.

He smiles. Dangerous smile. Mark seldom smiles. Sudden lurch forward — fist into Gerald's face. Gerald falls, crashes into a cobwebby central heating shaft. When he shows his face again it is glistening wet with blood and tears and sweat and saliva and self-pity.

— What are you going to do now?

Gerald lifts his hands, first to indicate bafflement and then to cover his face because he thinks Mark is about to punch him again when all Mark is doing is lifting the still-burning cigarette into the air and then dropping it to crush it under his boot.

— What are you going to do now? Answer me.

Horrible expression on Gerald's big bruised broken face. There's no fear on it any more. He's gone beyond fear into supplication. The horror of not knowing. Guilt at not being able to please Mark and, by association, we hope, we doubt, Charlie. Finally, he croaks out an answer.

— I. I'm sorry. *I don't know.*

— Give me your handkerchief.

Gerald looks uncertain.

— Yes. Give me your handkerchief.

Gerald nods, reaches into his right trouser pocket, out comes a silken red handkerchief and then, knotted to its end, a silken yellow handkerchief, and then a blue and then a green and then we're back to red again. He keeps pulling, both hands working like he's constructing a handkerchief ladder to climb up to heaven on.

— Enough. Wipe your face. Clean your face.

— *Thank you.*

Humble Gerald. Sincerely grateful Gerald. Mark is his angel and his world. Gerald looks at Mark as he works the handkerchief at the corners of his mouth and ears and eyes, and gingerly around the nose which still oozes blood and snot.

— I'll tell you what you're going to do.

Some vigorous headwork from Gerald. He's desperate to know.

— You're going to fuck off. You're going to keep out of our way.

Uncertain look from Gerald. Something odd has been punched and beaten into him. Mark the agent of violence is loved. Charlie the bystander is hated more than ever. Charlie sees this. Mark sees it too. It makes him sad. So sad he takes hold of Gerald's face between his hands and knees him in the groin and, as Gerald starts to crumple, holds him erect by the ears.

— Please, says Mark, don't make me repeat myself.

— I'm sorry.

— Tell us.

— Fuck off.

— Good. And?

— Keep out of the way.

— Very good.

Mark lets go of his face. Gerald falls against the wall, bounces forward, Mark holds him still with a hand tenderly against his chest.

— Here. Hold on to my arm. I'll help you out.

They walk, the three of them, with Gerald hiding his damaged face under a mask of coat collar and handkerchief. They climb the stairs to the lobby and Mark helps Gerald on his way out into the night with a gentle kick, as Charlie struggles with the boxes of candle sticks and miscellaneous kitchen items which had been his putative reason for getting the storeroom key in the first place.

Upstairs, in the apartment, Mark drinks mineral water and bathes his hands in a sink full of ice. Charlie uses Dinah's last will and testament as roach material.

— Where, asks grateful, awed, apprehensive Charlie (breaking his recently made vow for only the second time not to corrupt his mind and mission with narcotics, by taking a big grateful pull on the pure grass reefer), did you learn that kind of stuff?

— Films.

— What can I say? You're very good.

Mark wraps his right hand inside a Brighton beach tea towel (views of piers and Pavilion and dolphinarium).

— I'm off now.

What?

— No. Please. You can't. I have to meet Chatterjee.

Mark shakes his head.

— That's not for an hour or so, and you've got that woman coming first.

Of course. Rocky's girl. Maybe can persuade her to come along to Chatterjee. Bill it as an experience for her. The new Jews. Does this mean he can fire Mark now?

— Wait at least for her to get here? Yes?

Mark sits in the kitchenette. He occupies himself by using the car keys to scratch his name in the acrylic table top. Charlie snorts another line of coke. Then he takes the chair by the window and inspects the river. Over on the north side the police have stopped a motorist. Ant driver and ant passenger get out of car. Three beetle cops form a semi-circle around them. Looks like some violence might be in the offing.

— Where are my fucking binoculars?

The motorists are probably black. They usually are. Can't tell from here. And there's a light just switched on in Charlie's favourite window, a top-floor bedroom in which the action has been building up for weeks to the tentative coupling of a recent mother and her young homesick nanny.

— I had them, them, yesterday. Last week? Fuck it. Mark!

There's no answer. He feels the absence. The boy is gone. Charlie is alone. His heart beats wildly. Stars dazzle his eyes.

The doorbell rings. Thank God. Rocky's girl has arrived. Charlie gets up to let her in. What's the timetable after that? Chatterjee at the Happy Eater and from then, enriched, on to Scandinavia. He has everything he ever wanted and the day is going to turn out bad.

Charlie in Italian briefs and English string vest settles on the leather sofa.

— Come over here, honey.

Honey isn't making a move. Honey is leaning becomingly against the wall with her hands in the pockets of her baggy linen trousers. Honey is acting up even more than usual.

— Look. Come on. What's the fucking problem?

The problem is Rocky. Rocky is wound up. His fight is tonight. He has six hours to kill and he is going mental.

— Come here, honey. I can sort out your boy. What was I saying? Saying. Yeah. I know this, *therapist*. Massoose they call them. Inga. Properly trained, you know. Swedish massage, aromatherapy, everything, the works. Your boyfriend needs some calming? Sounds like he does. Send him to Inga. You got that? The place is called Scandinavia. I got the card somewhere. There. The Leviex bowl.

Rocky Sporgente's girlfriend deigns to extend an arm (and Charlie loves how the sleeve falls back, for a moment the fear recedes, slender pale wrist, golden down that catches the late afternoon river sun). Her hand moves to pluck the massage parlour's business card out of its place but it isn't there to be found — just taxi cab firms and food delivery services and house cleaners and professional girls who will come to your house and take plastic. Rocky Sporgente's girlfriend lets business cards scatter around her on the floor. She stretches. Arms up. Hips forward. Rocky Sporgente's girlfriend is not wearing underwear. That is cruel.

Charlie gets up. He walks towards her. He reaches out a hand to her, starts to unbutton her shirt. He thoughtlessly forgets to bruise her skin as he does so. She protests. He can't bear to hurt her. Not today. *Action*, he whispers, or maybe just thinks. A sudden move, down and around, and her shirt is open and her trousers are down and his chest is against her back (his mouth nibbles harmlessly at her shoulder) and his dick nuzzling up at her. He pushes her gently down now, something complains, her voice, his knees, and oh so smoothly he slides his dick into her warm lovely cunt. This is good, he wishes he'd thought of this before, his weight is upon her and she can't make him do anything bad. He's sliding in to her and away, and she is shouting at him. Violated by tenderness she's yelling, screaming at him, but he just keeps on going, fucking her from behind, mankind's oldest sport; he holds down her hands with his hands, his belly surrounds the base of her pale ridged spine, he's going to come in a fraction of a second, and he does, and as he does, with yelps and yeahs and rumbles that sound like barks, he sees, through the rigid clear polythene of the bookcase, beneath the Leviex bowl (imitation marble effect, milky plastic veined with green), a business card for Scandinavia.

She is squirming under him, fighting to push off his gentle weight, to turn it into something aggressive, and oh, and as the last few shots of jissom fire inside her, as the presentiment hits Charlie that he will never fuck her again, he finds her wrists to nail her down, the gentlest crucifixion, to the floor.

She will never forgive him. She takes the Scandinavia business card and leaves. She refuses to go with him to meet Chatterjee, despite his pleas and despite his (I am ashamed to reveal this) tears.

TERROR VISION JOURNEY. The world is like this: smoky, shiny, scary, about to collapse upon itself in every moment and only he sees this, why? The sky to tumble upon our heads. How can those cars stay in straight lines like that, it's an effort of group will that is likely to splinter any second, a moment's absence and they'll all be crashed, wheel-spinning, dead. The drivers make their consensus decision to stick to the rules, to steer between the broken white lines, but in every group there's always one troublemaker at least, more than one — is that him? the shifty-looking bastard turning the wheel of his convertible — or that one? in the company sedan, you can see it in his eyes, drunk on liquor and suicide — or is it Charlie himself, is he the rotten one here, the one who will just pull the wheel across into the sickening mesmerising lure of collision . . .? Charlie tries the radio and that only makes things worse, every voice snidely mocks him, each snatch of music is the soundtrack of imminent death. He keeps the car in third and drives with one hand on the wheel and the other against his failing heart. He makes it to the car park at LPC. He abandons the Scimitar there, walks on to the Happy Eater.

The paving stones have been laid unevenly, or maybe it's the pressure of chaos from below that pushes them into this shape. Charlie looks into the cars going past and he can see bones beneath flesh, worms beneath bones, all of it breaking up, churning, rotten. His legs keep going, must not think about them, if he considered his feet then walking would become impossible, like driving, like speech, like connection. His poor heart, spluttering out its beats, threatened on all sides, by cholesterol, by fumes, by cocaine, by tar, by life itself, it keeps going, gamely — but don't think about it; that was a lurch for certain, an absence where the failing beat should have been: ba-duh, ba-*duh*, ba-*??!* Open eyes, keep going. Look for comfort. That hedge there, optimistically planted, wilting, winter brown, a boarded-up house behind it, visible through the dying twigs. Think of the sun behind the clouds. Think of the fortunes about to sashay into his bank accounts. Think of his dick, indomitable. (Even when things have been at their worst that at least has never let him down, or hardly ever.) Think of cocaine. Think of cannabis. Think of Conchita and her sisters. Think of plastic. Think of freedom.

Even the Happy Eater fails to console him. He walks in, pretends to be alive, and as he sits he tries to ignore the bankruptcy in everybody's eyes.

Where is Chatterjee? Not like him to be late. They've waited longer than two years for this. Fuck him. He will be here soon enough. Everything almost is in its place. Charlie reaches up to tilt his panama hat. He has forgotten to wear it. He has no hat to tilt. Charlie starts to whistle an Irish song his father used to sing.

— Sir? Are you ready to order?

— A moment. Oh fuck it — excuse me, er . . . Tina — I was miles away. Just get me numbers ninety to ninety-five.

— Are you sure? That's–

— Whatever. Inclusive. And a beer. Lager. Pint.

— We only serve half-pints, I'm afraid.

— Then I'll have two of those. That's lovely. Off you go.

Charlie breathes deeply. *Come on.* The world is being made into Charlie's chosen shape, and Chatterjee is marching into the restaurant, the future is here. The modern moment has arrived. As they say on the menu, *Enjoy!*

Charlie goes through the checklist with Chatterjee. His mind is not entirely on the job. Everything feels wrong.

— Mr Levy. Mr Levy? Please.

— I'm sorry, son. Truly. Talk to me.

Even this place is ruined by precariousness. He can take little joy in being here any more, he finds no comfort in the polyester silk plants and decorative laminate walls.

— Mr Levy.

— Yes. I know. Hold on a mo. My order.

Charlie pretends to listen as his peculiarly shaped waitress (thick neck and thin torso, skinny hips, thick legs) delivers a series of plates to his table, most of them containing fried food coloured orange, with some breaded mushrooms on the side. Normally this would delight Charlie. Normally he would be listening to the jingle of food names. Maybe, he decides, it is the enormity of his success that is unsettling him. A signal of the inevitable change. He must be strong.

— Sign, I believe, here and here and here.

Charlie signs. Passes back with the papers the details of the bank accounts into which the money will be paid. (He keeps the pen, Chatterjee's own, a thin added victory but it consoles him all the same.) He should be celebrating. He should be reading the small print. He should have done this before. He should never have done this at all. The walk back from the Happy Eater to the factory frightens him in its lonely, traffic-surrounded prospect. So many opportunities for his enemies to strike. He has no notion of who his enemies are, just knows they are mounting, Charlie's legendary luck about to give out at last, faceless malevolent hordes.

— Excuse me. Do you feel unwell?

— Never better. Fancy a drink? We should be celebrating.

His face feels like it is breaking in the effort to keep a smile on board. He remembers to breathe. Mr Chatterjee refuses a drink, performs an elegant farewell and leaves. Charlie pushes aside the wreckage of his meal, touches, for luck, the photograph, and slowly starts on the difficult solitary walk back to the car park. Mark has abandoned him. Rocky's girl will never forgive him for defiling her. His war with Howard is over. There is still Leslie to punish. Think of that and think of Inga. She'll look after him.

4

OH SCANDINAVIA. HEART'S ease. Ooh that's nice, that's the spot, yes yes, right there, oh my, those clever Swedish fingers, keep going, don't stop, thank you, Inga.

Charlie naked on the pine bench, a hole for his face to go, Inga's fingers working away now at the acid deposits that knot around his shoulders.

— Woah! That hurts!

— It's meant to, Mr Charlie. Don't be big baby.

— Why do you talk like this the whole time? Ow! Fuck it! Like you're some Thai-stick whore? I know you're a fucking university graduate. Umm. Yes. That's cleared it. That's nice. That's *nice*.

Geranium and basil oils rubbed into his skin. His body is draped in towels. She will work down now, each little vertebra receives its due, a couple of slaps of his bottom, then down his legs, lifting up the next towel, a little close reflexology work with his feet, and then over he goes, and her fingers are up to his face; and it's always when she starts her odd little shivery massage of the face that he drifts off, when words run out and he finds a peaceful place to stay.

— What's the basil for, Inga?

— Manly smell. Big boy Charlie. It is the geranium that works upon you. Basil just for balance. Could use rose or lavender oil, much the same effect, different odour. Mr Charlie. No stress, bad boy.

— You ought to have your own place. Call it Inga's Fingers. A finger of Inga's goes in just a little too hard.

— Hey!

— Don't tease. I dream of place. Respectable. You think I

like it here? You think I send postcards home — oops, excuse me, missed — dear everybody I work in sex fuck knocking shop, happy Malmo summers? Give me a break.

— Maybe I will.

Dreamy. Here at least there is assurance.

— Turn over big boy.

Big boy turns over, a wave of flesh, his big boy dick (which is not, let's admit it, long, but the width is what pleases the ladies), yawns from one thigh to the other, ready to be called upon if necessary. The towels are rearranged, his body is covered, she's going to rub oil into his hair and then pull clumps of it about, and then she'll do the shivery thing with his face.

— I tell you, I'm going to drift off in a minute but you know, Inga, things have come my way.

Charlie the dispenser of just charity. To Inga an aromatherapy centre.

— Soothe down, Daddy. No stress. Easy now, big boy.

Maybe Inga should be his chosen companion. She's too tall and too blonde and too old but he loves her.

— And hey, stud. Give me break, huh? Darling? No more drugs. The state of your heart not so good. I see brown specks in your eyes.

— And in yours, honey. Sure, sweetheart. No drugs. I mean it. I want you to have what you want. I'll take care of the lease. You'll have to pay the service charges though. Inga's Fingers. You like the name?

And then she starts to do the shivery thing with his face, ghostly, her hands slipping up and down his cheeks, twitching. Charlie drifts away, into that dark brown place she sends him to that tastes of mother.

He is protected by Inga's oils and fingers. A force field, like the one he used to dream about as a child, surrounds him. He can come to no harm. He is impregnable. He is glorious. Out through the sex corridors of Scandinavia, out past the reception area, a quick wink at the wiry little Italian in shades and leather jacket coming in. (Hey that's Rocky, that's the boy himself.) And quickly now, let's not blow it, let's not take too much for granted, we're multi-millionaires now but there's still that fear ready to strike — into the car, away from the kerb, one more stop back at the factory to pick up the drugs in the desk, then it's the drive home as quick as we can, with a force field scented with geranium and basil surrounding the car.

5

ROCKY'S GIRL HAS left a message on the answer machine. It is the first time she has done it and, we can be sure, the last. As he listens, Charlie removes his clothes, wraps his shivering sweating body in towels, and tears off months from his calendar.

The first part of her message is entirely heavy breathing but the intention is not arousal. Eventually she remembers to speak.

— Yeah? Hello? Who is this? Hello? I'll hang up, fucker. You ready for this? —

And the most piercing whistle shrieks out of his machine. Charlie fears for his neighbours, his walls, the foundations, himself. He covers his ears with his hands and damp towels fall to the tiled floor. After the whistle dies comes a silence and then a laugh, like something delicate being smashed.

— Sorry, Charlie. I forgot I was on the phone. I called you, didn't I? You all right? I've never used that before, good isn't it? Police issue. I'm calling you because, even though I hate you, because — Charlie . . .

Her voice emerges naked from the machine. Stripped down to its pretty girl core, unadorned.

— I don't know what's going to happen now. He hit me, Charlie. He's, I don't blame him, he's terrified. It's your fault. He went to that massage place you said, saw that woman Ingrid —

— No! Charlie yells, I said *Inga, Inga*. Not dirty *Ingrid*, we had her a couple of times, didn't we? not the right sort of girl for Rocky at all. *Inga*!

— And I don't know, he fell asleep or something, whatever. It's hard to work out what happened, he's in a state, 'sfar as I can tell she jerked him off, he lost his precious seed. Oh Charlie, he's devastated, he's so unhappy, it breaks my heart, it does. Poor Rocky. He's fighting tonight, it's on the telly and everything, and he's lost it completely, that whole self-belief thing. I can't bear it.

That's the extent of the message. She doesn't hang up for a while, she's too busy contemplating the ruins of her boyfriend's boxing career, then she remembers finally to put down the receiver. There's enough space after that for the start of one more message, a crowing thing from Tessa inviting him to some event but fortunately the tape finishes before she does. Charlie turns on the TV and waits, only interrupting his vigil to fetch more vodka and cigars and snort up more cocaine, for Rocky.

The boy is in a state. That much is plain. The cocky Italian strut is gone. He looks lacklustre, half-asleep. His trainer, a short guy with bristly white hair, sits him on his stool and slaps his cheeks and whispers things and kisses him. The opponent looks like a fairground boxer, hairy chest, balding, tattoos. Charlie espies Rocky's girl in the third row holding a handkerchief up to her eyes. Charlie experiences a brief flush of pride when he recognises the handkerchief as one of his. This is to be Rocky's warm-up fight for the big one in the States. It's the undercard of some European championship bout, and all Rocky has to do is look good and beat the other guy too. But Rocky, his gold crucifix disconsolately dangling over his trainer's wrist, trudges from his corner with his gloves held low. In the centre of the ring he sadly pretends to listen to the referee's instructions. Rocky does not look as if he could give Dinah's corpse much of a fight.

Rocky crosses himself, taps gloves with his opponent and even that seems to hurt him. He winces. The loss of his seed has unmanned him. His fairground opponent rattles around the ring, pushing clumsy jabs at Rocky, sometimes winding up for the big right hand punch. Rocky backs away. He still looks graceful, even in retreat he has all the fancy moves, but he looks like a loser all the same. He pushes out at the mauler with gloves together, he ducks away from the jabs, but regretfully, as if he doesn't deserve to escape. The mauler's coach parties of tattooed chums and incestuous cousins begin to holler and chant. The commentator, with a teary catch in his voice, celebrates the mauler as the latest great Englishman.

Rocky spends the last part of the first round ducking punches on the ropes. The mauler is so keen to land his knockout blow (in one brief shot the cameraman goes in close on the mauler's piggy, bloodlusty eyes) that he neglects to pin Rocky properly down. The blows are parried by Rocky's gloves, they graze Rocky's shoulders, skim over the top of Rocky's balding bobbing head. The bell sounds. Despermed Rocky has survived one round at least. His trainer slaps him on the cheeks. Some of the crowd jeer at Rocky, the rest (apart from one frightened girl) cheer the mauler.

Rocky starts the second doing his best. He dances on dainty feet, he jabs the mauler, he pushes him back some paces, but his heart is not in it, that's plain to see. He has been undone by Fate. He catches the mauler with a good one all the same — a left cross to the side of the mauler's pocked face — and the mauler shakes his head and the crowd is yelling at Rocky to finish the job but instead he retreats and the moment is past. The mauler, who looks like he can't believe his luck, gets Rocky on to the ropes again, and this time, with his knees bent to give him better leverage, he forces a big punch through. Launched from next to the mauler's right shoulder, it goes under Rocky's left glove, over his right, it cracks into Rocky's unsurprised chin. Rocky steps forward, the mauler misses him with his follow-up left, then Rocky sways backward, and Rocky goes down.

Poor Rocky. (And poor Rocky's girl, whose head is on the shoulder of Rocky's manager.) He lies there, his gum shield half out, his hands on the floor. The mauler jumps up and down as the referee guides him into a corner. Rocky looks around. At the ceiling (decorated, he might notice, with cherubim worked into the moulding and cornices), at the crowd around him, and, by fortuitous means of the television camera, directly at Charlie. Rocky's gaze is clear. His mind, which is on the slow side, is trying to work something out. Charlie, like most of the crowd, gets up to his feet.

— Get up you wanker! Get up! You're not fucking hurt! He hasn't knocked you out!

It is true. Rocky takes an age though to realise this. The punch was strong, it knocked him off his feet, but that was all. It may well be the first good punch he has ever taken in the whole of his career that left his senses intact. It's the shock of this, not the physical effect of the punch itself, that is keeping him on the ground. The referee is counting. The mauler's hordes are counting ahead of him. Rocky's girl is sobbing on the manager's shoulder. Rocky's manager has insinuated a sly comforting hand underneath her silk shirt. And Rocky is shaking his head. And Rocky is lifting himself up by pushing down with his elbows. The referee has got to seven in his count. Rocky's stiff torso, like a Boris Karloff monster, is rising. The count is eight. And Rocky, the acrobat, pushes down with his weight and flips up into the air and lands on his feet, sways, gathers, pushes his mouthpiece back in, nods to the referee, and waits, senses unscrambled, for battle to recommence.

It is magical to see. Charlie bellows himself hoarse yelling Rocky on. The crowd, most of it, is on his side now. Rocky's girl has disentangled herself from the manager. But Rocky doesn't need any of them. He is filled with a simple, thoughtless power. For the first time in his life Rocky has been hit hard and Rocky has not been knocked out. There will be time enough to consider this later, with his girl perhaps, or his manager, or his mother, or with the flatterers of the press, or the toadies of the boxing association. Now is for action. Now is for show. And Rocky puts on a show, for the watching Americans, for his girl, for himself. Rocky jabs and weaves and dances and hits. Rocky smiles, there's a new, harder Rocky now, we each of us find our own moment to become adult: with each punch that he lands, his gloves twist in a little at the moment of impact and the mauler's skin opens with every one. For the rest of the round Rocky shows off his skills and shows up his opponent, and then, with seconds to go, Rocky lands his own big punch. A jarring left, then a reconnoitring right, another jarring left, and then this sweet right cross, delivered with love, lands on the side of the mauler's face, and puts the mauler to sleep.

Charlie is cheering. And Rocky is jumping up and down. His trainer is weeping and hugging his boy like he's just won the championship itself. There's a man with a microphone trying to beat a path to Rocky through the suddenly packed ring. And there's the mauler, still out cold, perhaps brain-damaged, we'll never know. And as Charlie feels himself tugged down by the sudden, quite painful need to sleep, there's Rocky's girl, face coarsened by tears, face more beautiful than ever before, hugging her triumphant man.

NINE

THE FUTURE

I

TOGETHER HOWARD AND Mark search through the drawers in Charlie's kitchen.

— He keeps another set somewhere here. They say Scimitar on them.

Cutlery and napkins and photographs and correspondence in elastic-banded bundles. Balloons, candles, condoms, milkshake straws cut in half. Packets of rolling papers missing their covers. Books of matches from fancy restaurants. Detritus from a half-completed life.

— Not here.

Howard goes back into the living room where naked Charlie sleeps. Mark has covered him with a blanket from the bedroom. The toes of his splayed feet peep out at one end, his sleeping face from the other. The enemy's eyes are caverns, his nose and hair mountainously huge, trick-illuminated by the muted TV screen, by the lights of the apartment buildings on the Chelsea side of the river, by the moon hanging low. On the mirror built into the acrylic coffee table are a rolled-up £20 note, an opened envelope of paper, and a line of cocaine waiting for Charlie to attend to it. The photograph is here too, propped up on the table. Howard quickly checks that Mark has not followed him in. He slips the photograph into his jacket pocket.

Howard squeezes in to sit beside Charlie on the sofa. He's not moving, he's silent, not even a snore, he couldn't be . . .? No. Impossible. Not Charlie. Howard pushes aside a cigar butt, a calculator, a bag of grass, a business card for a massage parlour. It is hard not to smile, the nude buffoon beside him about to lose his early morning wake-up shot.

— Oi!

375

Mark has followed him in. Mark doesn't like what he's doing. Howard won't be stopped.

— Sssh, don't wake him. I can't resist. You realise that this is the first time I've ever got anything off him for free?

Charlie certainly does not stir, the silent ugly sleeper. Gleeful Howard leans forward, introduces the banknote to his left nostril and snorts up the line. He wipes the note clean and drops it back on top of the mirror. This — he winces at the acrid taste of cocainey saliva dripping to the back of his throat: he smiles as the drug rushes up to his brain to inform him he is a godling — this is going to be good.

— I found the keys.

Mark pushes past him, lifts the blanket from where it had slipped down onto Charlie's cold shoulders, a pause, then he moves it further up to cover the face.

— What are you doing that for?

— I said I found the keys. Let's go.

The photograph is getting dangerously tatty now. Blanched like the pale perfect legs of Rocky Sporgente's girlfriend. (But how could Howard know about her?) Sticky like the handkerchief inside Leslie Singer's bedsheets which he leaves, tell-tale debris, where it lies after his midnight masturbation sessions, but still, even after this much time has passed since his mother's death, he gets up, naked, a silent hard-breathing performer shivering in the unheated air, to replace the vaseline jar in the bathroom medicine cabinet and the ethnic photography book to the glass-fronted library shelves in the parlour. (And how could Howard know about that? Well, he doesn't. There's something else, a force, a presence, a tattered connection over the abyss between lost past and plastic future along the narrow bridge of the uncertain present.)

They had met on the Heath. Near a bench overlooking the swimmers' ponds. Howard had gone there thoughtlessly, purposelessly, just walked. The shiva was still on at his house, Dinah still being mourned by friends and relations who had not yet forgotten, in their grief, how to eat. Howard had stood on the side watching Ruthie flirt with Andy Astaire. He had tried to think of something to say to his daughter, something that would arm him to go across and pull her to him, but short of the truth he couldn't think of anything to say. Howard Levy, jobless, motherless, suddenly sonless, always fatherless, quit the house. He walked away, out of the front door, a zigzag path around parked cars and off through the opened electric gates. Opposite, a light burned in Miles Rodgers's living room. The bald crown of Rodgers's nodding head was just visible over the top of a sofa as he plotted future patios, no doubt, to the soundtrack of some Broadway musical. No purple light shone behind Simon's window.

Slowly, Howard walked away. Unfamiliar familiar smells. Car fumes. Dog shit. Sharp autumnal air. He had to force himself to walk slowly. The urge to run lifted up inside him and it was hard to resist, not to run away, again, evade them all, for ever, maybe to complete Jack Levy's original journey and then find a Cornish barmaid adrift in New York, or to go somewhere unlikely — Maurice could be his guide to places where they talk comical English and the girls will do anything for a dollar. A dog barked in an emptily lit house. A car's headlights dimmed and died in the next gravelly drive. Two pairs of blonde schoolchildren kissed under cover of a hedge. Horse-head pillars. An elderly woman was passed by two burly types from a taxi cab onto the wheelchair ramp of a private ambulance.

Howard carried on up the street to the beginning of the Heath. The sky was good and so was the cold. He kept walking. He found a bench to sit on to observe the deserted ponds below. On the back-rest of the bench was a metal plaque that the night made impossible to read. (Didn't Leslie endow a bench for Freda? Didn't Danny and Dinah perform some of the moves in their awkward courtship here?) There were stars and he was shivering and he felt that he could have sat like this for ever.

A boy was sitting not far from him. Perched up on the back of an opposing bench with his feet on the seat. He looked as if he was waiting for something, for somebody. A rent boy waiting for the unsheathed desire of an older man. A mugger, a killer, waiting for a fool to sit out at night beneath the stars. The boy climbed off the bench. With his hands in his pockets he approached Howard, the calm walk of the professional badman. Howard wanted to run. He had done too much running already. The boy sat beside him. Howard huddled into himself. The boy jogged him on the elbow. Howard did not want to establish eye contact but he did so anyway.

— Oh. It's you.

— I'm impressed.

— Excuse me?

— I didn't think you knew about here.

— Knew? Knew what?

The almost admiring expression on Mark's face was replaced by the one that Howard had associated with him ever since their odd conversation at his welcome home party, a guardedness touched with scorn.

— Where he comes. To sit. Your son and heir. The bench over there that I was on. Just sits. Solo. Nothing to touch him here.

— He's gone. Run away.

Mark shook his head. Superior wisdom.

— He has. He's run. I broke into his room. No one there. I thought, maybe he hasn't been there for years.

— He does that. Goes for walks at night. He likes his exercise. Climbs out over the roof and the side of the house. Once I saw him doing it when I was . . .

— When you were what?

— When I was fucking your daughter actually. If it's all the same to you. But you weren't around so you can't talk. That's what he does, climbs out, climbs away, wanders the streets for a while or sits in the park. He always comes back. Homesick, the prince. Climbs back in through his window —

— I closed his window.

— You fuckwit.

— How was I to know? It was cold in there. Blustery. I shut his window.

— Think of him. He gets upset. Where's your car?

— Not here. And I don't want to — go back to — It's unavailable. But how do you know where to look for him?

— I don't. Let's go to Charlie's. We'll pick up his car.

They caught a cab to Charlie's place where Howard felt

joyously like a thief picking through the glistening rubble of his cousin's life. The bedroom was a disappointment though, the master cocksman's lair. CD player and monitor in the wardrobe with a pile of music and movies, some pornography, some not. A stack of books, real impress the ladies stuff with eastern poetry and western business philosophy, on the bedside table. Five little brown jars labelled with the name of a different fruit or herb or spice. (Frankincense. Tea tree. Grapefruit and basil. Lavender. Geranium. What are they — love potions? erectility serums?) An industrial-sized pack of condoms beneath the bed. Howard had been tempted to unwrap each one, make a prick in each and watch the horrified world be polluted with tiny Charlies. The boy was a strange ally. He had some feeling for Simon but wouldn't say what it was.

— I said I found the keys. Come on. Let's go.

Mark is, Howard suddenly realises, slowed by sorrow. Even his impatience is somehow softened, made ponderous, dignified. Who is he grieving for? Simon? Himself?

— Why is he called Porky?

— Eh?

— I thought you might know the answer. It's a joke. It doesn't matter. I don't want to drive. I want to walk.

— Please yourself.

Off they go, Simon's pursuers, and when they get to the car park outside capricious Howard keeps walking and sad Mark follows slowly in the Scimitar.

2

LONDON IS FOR nighttime. Howard walks east between the river and the road and the ordinary fug is pushed aside. A man freewheels past on a bicycle, humming his exhilaration of speed. A lorry lumbers by, the driver grinning, could be a joke he's just heard on the radio or maybe it's something he's thought of himself which he's polishing in his mind like a jewel to offer to the lover waiting for him. Cocaine is driving steel roads through Howard's brain. London is beautiful. This city he shares with lovers who kiss on bridges, in telephone booths, in parked cars; with solitary madmen who must walk alone, overcoats flapping; with children sharing sleeping bags in shop doorways; with drunks who by this time of night have safely neutered themselves with liquor. The rain comes down in a drizzly shower, all of them here on the streets are cold and wet equally, together, haloed by the headlamps of cars; and as he starts to cross the river, with the Scimitar keeping to a steady dawdle beside him, he is momentarily dazzled by a flash of heaven behind Waterloo Bridge scaffolding, inexplicable, perfect, and gone.

3

SOAKED HOWARD, RAIN-VOLUPTUOUS Howard, consents finally to ride beside Mark in the Scimitar. He flags him down on Kingsway, sits squelching inside. Mark is listening to some ugly music, machine sounds like metal insects making robot mating calls.

— Where am I driving to?

— I don't know.

Howard tries to light a cigarette but the pack is far too wet. He replaces it in a soaking pocket. Takes out the photo hoping it might dry on the dashboard.

— What's the connection between you?

Mark is silent. Annoying.

— Are you friends? Or what?

Mark steers with his right hand and chews on the nail of his left little finger. Lights flicker across his face. His curly black hair is pulled back in a long pony tail. His eyes are narrowed. His cheekbones are high. He looks like an Irish prince.

— I used to tell him things, for nobody else. I would have words for him now. I used to sit with him sometimes.

— Sit with him? Where? At the house you mean.

Mark switches hands. He steers with his left hand and chews the nails on his right.

— In the park most often. He would sometimes talk to me then.

— Talk? What's his voice like?

Mark looks at him. It wasn't the expected question. He shrugs.

— Watch out. Dog.

Its eyes are yellow, fearful, somebody's missing mutt. Mark

swerves the car around the dog, a taxicab brakes, a cyclist wobbles, a pedestrian falls against a rubbish bin and shakes his fist. Mark pays no attention to the ensuing abuse of words and horns and lights.

— Voice? Pretty. Songful. He feels things.

— Things?

— He prefers numbers better. He says numbers are safe and nice and have fixed characters not like people. He used to have this ambition to count up to a million, personally. Say every number out loud, no cheating. I don't think he's got there yet. He's always being interrupted by people telling him things. He got hooked on that. People's little secrets. Power and glory and dirty little stories. But it weren't just the words. He listens to, can hear, feels, smells, what goes on inside. Too painful. Hurts. Where to?

— I don't know. Just keep going. What direction are we heading?

— North. Before that it was east.

— Let's do west now. Isn't that the route that discoverers take?

— And he's hopeless too. Dead clumsy. His clothes are always torn from where he's fallen over or banged into things. Like an idiot. Although he did fix my Walkman once. In the shop they told me to throw it away. That American, Yank-wank Arnie, he had a go and made it worse. Didn't play at all after he got his hands on it. Simon fixed it for me. I'll give him that. Where are we going? You've worked out where we're going haven't you?

— Keep going.

— I used to tell Charlie sometimes to give it a go. Just go upstairs and speak. It would have been better than whoring and scoring the whole time. It might have done him some good. Poor Charlie.

Howard is meant to concur in that. Epitaph for an ancient enemy. Why poor Charlie? Every time Howard's thoughts approach the image of Charlie something makes them veer away with a lurch of nausea and pride. It is as if Charlie is surrounded by a force field that protects him from Howard's mind and even memory.

They drive along Western Avenue, past the Hoover building and past signs to the Polish cemetery and past where the lamps are dwarfishly low because of the RAF airport.

— Just here. After the Happy Eater.

— I know where the factory is. I'm an employee. You fucking hired me.

Howard reaches up into the night light casing and removes the key. He unlocks the door. Mark pulls it open. They each wait for the other to go in first.

— Listen, says Mark.

Yes. Listen. If we block out the sounds of the cars racing each other along the banking curve of Western Avenue, and if we ignore the buzzing of the loose connection to the halogen night light and ignore the heavy weight of our own breathing, then the noises might become clearer, perhaps even interpretable.

— Sssh. Listen.

Push away all the surface noises, the accidental city music, and what's that underneath, clanking, insistently slow, almost regular? Inside the factory somewhere metal grinds secretly against metal. Machine fornication, illicit, unlubricated. A slow metallic fuck. And then there's also a whirring, an industrial whining, and above it all is the top-note screech of a mechanical banshee. So this is what machinery does when men aren't around to stop it.

— Come on.

It doesn't feel right to go in, it would be a sort of intrusion.

— Perhaps we should just . . .?

— I said come on.

When Howard consents to start walking of his own accord Mark lets go of his sleeve. They go in darkness along the corridor. The noises get louder. Ahead of them a stream of pale light leaks onto the corridor from the factory floor.

Maybe there will be a ballet in there, accompaniment provided by an orchestra that uses machine parts for instruments, Charlie the bandleader dragging out that relentless ugly rhythm, while half-human half-mechanical puppets vitalised by electricity dance an evil choreography. Howard and Mark climb the three concrete stairs. Mark pushes open the door.

— Careful. (A warning hand from Mark on Howard's arm.) He scares.

— *He* scares? What about me?

The door's wheeze as it closes behind them is almost unhearable in the din. Museum Corner is alight. Howard and Mark walk across the shop floor, fewer machines to walk past than last time. There they are, Rosie and Christine and Bertha and Ruth and Agnes. The machines have moved or been moved into a circle. They face out, proudly, and all are aglow, alive and awork, tethered with electrical cables, fed by cylinders. Jerking in awkward repetition. Given life. Among them, beside the screwdrivers and the scraps of cable and the bags of screws and the thin electrical wire and the sampler-sized drum of plastic granulate (fetched we suspect by the forklift truck that Mark used to drive and which is now nowhere to be seen), there should be a jar of mother-made pickled cucumbers and a boy brushing the flop of dark hair away from his eyes. But little Sarah Bremen is dead and Danny is dead and Vivien does not make pickled cucumbers and anyway Simon is not here.

— It must have been Simon. Mustn't it?

— Maybe the tooth fairy. What do you think?

What does Howard think? Howard dares to think nothing. Slowly he enters the charmed circle of ancient, restored machinery, glowing happily industriously erratically red. All of Danny's useless machines are switched on. Simon has understood them, as Dinah knew he would or maybe just hoped. Cables entwine, plungers shudder, wheel valves spin, die plates kiss, sprue brushes wipe, and he realises where everyone had gone wrong before. These are not isolated machines. They connect. A system with five parts. Rosie heats. Bertha injects. Ruth moulds. Agnes adds a plastic wrapping. Christine applies the hair. (He may have a knack but the boy is no genius — probably better to add the hair before the plastic wrap.) And out comes the product, a slow, monstrously ungainly operation, but here comes another one, belched out of Ruth, complete, cooled, hunchbacked and spattered, a little bald plastic woman, perfectly pretty and anatomically, if eccentrically, detailed, waiting for Agnes to gift wrap, and for Christine to apply the most lavish red hair to the outside of the package.

Rosie's screws are working out of her thread holes, spinning up thin ribbons of steel. Here comes a plump new doll ejected from Ruth, wrapped by Christine, hairdressed by Agnes, playfully rolling down the chute into the collection basket. Already, just like LPC only speeded up, decay has set in. The first doll, despite having a vagina at her hip, is really rather attractive. There is something appealingly goblin-like about her dimpled features. She is almost, Howard has to admit, sexy. But the moulding plates of Ruth are refusing quite to meet. Each subsequent doll is more smearily featured, limbs more lumpen. Soon the plastic will just congeal shapelessly and it will be impossible to guess at what was in the inventor's mind or heart. Howard finds the breast plate of each machine and switches the power off. One by one they are deprived of electricity and left to cool, sulky, panting. The last dollop of hair dribbles out of Christine.

Beside red gleaming Rosie, Mark lifts up a man's torn shirt. Howard takes it.

— That used to be mine. Vivien made me get rid of it. There were some, you can see them, grease marks on one of the elbows.

— There's worse than that now.

The shirt is now ripped on both its sleeves from where its clumsy wearer has scraped against a tree or a gratefully working machine. Howard takes off his jacket. He slips the shirt on. This is maybe the closest he can get here to his son.

— Simon!!

Howard's voice is louder than he had intended. It startles them both.

— Simon!

Howard's voice is better this time. Measured. Full. Fatherly. Now here is something that is indisputably Howard's own.

Howard's voice echoes around the factory. A screw works its way out of Rosie, chinks to the floor. Silently, holding their breaths, they listen to the muffled sound of traffic on Western Avenue. Mark gets tired of that first. He says something about

Charlie and something about arrangements. Howard offers him a gift-wrapped doll to carry away, a last party favour.

4

— MY FATHER, SAYS Howard, died in shame, by fire, with a red-haired girl. We too might succumb to red-haired girls.

— Terrible thing, says Leslie, who shows the courtesy to tut-tut tragically. It makes him appear to be sucking on invisible chicken bones.

Leslie has been ennobled. Leslie is in charge. Leslie is sovereign steward of LPC. His regency duties may be few, to tot — and tut — up the figures, to finalise the accounts, to gather the assets into one legal place and supervise their handing over to Wiener, but Leslie none the less is blooming in the autumnal sunlight of interregnal power.

— Howard. I have to say. Very interesting. Tut tut. But you, excuse me, look terrible.

Of course he looks terrible. Howard has spent the night in his old office. He had tried the chair at full-recline and then at half. He had ended up sleeping, if he slept at all, on the floor. He had watched the morning approach through his office window. And then when it seemed light enough to call home and ask if Simon had returned he found that the telephone had already been disconnected.

Leslie's office is almost unrecognisable. It has been dusted and cleaned for probably the first time since Freda died. It has been decorated with photographs. There's one of Freda in her mannish business suit. There're Freda and Leslie enjoying a contemplative moment together over a desk covered with accounts. Freda and Leslie and Bertha on the water's edge at some South Coast resort. Leslie and Bertha as teenagers, on stoical ponies, trekking up a wooded path. There's Leslie with a woman unknown to Howard but the setting looks Scottish.

And there's Charlie, many of Charlie, at work, at play, behind desks, on the shop floor, on the tennis court, at a karaoke machine in a nightclub, surrounded by girls with dyed blonde hair at a restaurant table, and there he is at Horner's Hall shaking hands with different men who all seem almost famous. And there, in a silver frame, on the centre of Leslie's brave new desk, who knows how he found this one, an old shot of Vivien taken on a family holiday but the half of the picture that a younger Howard used to fill has been ruthlessly cut away. Is there no end to this man's ambition?

— How is the changeover going?

— It goes. I have to tell you a lot of work is what it entails. To do.

— Done a nice job with the old place, hardly recognise it. Is that a photograph of Vivien there?

— Ah. Yes. Not an impertinence I hope. Merely to give the correct impression for visiting executives.

— Get a lot of those do you? Sorry. I don't mean to jeer.

Nor should he. It is not that he is here for a favour but it is probably not a good idea to lose Leslie's good will before he has left with what he came here for.

— I'm sure you're doing a marvellous job. (Leslie preens.) Far better than I would have done. (Leslie nods, perhaps more emphatically than he might.) And I know you're very busy. (Leslie sighs.) I'm here just because Charlie and I—

Leslie dabs an alertly moist eye.

— A tragedy. He was twelve years younger than myself.

— I'm sure he was. It hits us all. It's just we had an arrangement . . .

— Really? Did you?

— It concerned some machinery.

— I cannot be, I am afraid, party to Charlie in his — saddest of all reasons why — absence. You would hardly think to hold me to somebody else's arrangement . . .? Not even Charlie would have expected such a thing. How we all loved him. I think even when he played the rascal there was good in his heart.

— I'm sure there was. How we all . . . But I agreed for the sale to go ahead and in return —

— The sale has I think gone ahead. Voted on, in person or absentia by all qualified to do so. Attested to its unimpeachable correctness by our legal representation. All above board. Ship shape and Bristol fashion.

— I don't doubt it. But Charlie and I had an arrangement. I waived my share of the proceeds, left it to the two of you, and in return—

— You and Charlie. You and Charlie. I keep hearing you and Charlie. I was not party to any arrangement.

— I know you weren't. That's not the point. Charlie said he would clear it with you. You weren't forgotten. We didn't go behind your back, anything like that. I had assumed that Charlie would have cleared it with you, got your agreement. It's, it's Danny, my father's, his old machines, sentimental value . . .

— Assumed? The hare assumed he would beat the tortoise. I did not give any such agreement. The machines belong to Harry Wiener now.

Howard is on the point of crying. His fists are balled up. He just wants to say *But he said . . .!* and let the tears come. Come on, shape up, a child could deal with Leslie, this blinking, nose-running man who has his hands clasped over the beginning of a tycoonish paunch. What Howard must do, to move this indomitably stubborn man, is to charm him, con him, like Charlie would.

— What do you want Leslie? What is that you want more than anything . . .?

Leslie leans over the desk. His muddy little eyes are sharp for once, malice has focused them.

— I have what I want. Thank you very much. Now. If you excuse me. I really am most busy.

389

THE LAND BELONGS to Chatterjee's people. The demolition men are already here, hard-hatted teams kicking through broken glass and wooden pallets in the factory yard, smoking cigarettes, chatting in groups. They're sizing up the task ahead, the equipment needed to tarmac an access road to the car park of shopping paradise. Paradise announces itself on a billboard high enough to reach into the future, and underneath, in even bigger letters, are the names of all the construction companies and management firms involved in the project.

And here comes Wiener: a big grey Jaguar pulls slowly down the lane, glides to a halt. The old man gets out, adjusts his hat, flicks a hand to brush off the help of his driver, a young black man with white robes peeping out from under his uniform coat. Sleepy unshaven Howard goes to wait on the shop floor. The only other person in the building is Leslie, putting the last touches to his stewarding tasks, probably enjoying the delicious taste of half a million pounds. Howard's eyes adjust to the darkness. The factory clock has stopped long ago at five minutes to five, almost knocking-off time, permanently. He stands as far away from Museum Corner as he judges appropriate. The metal door swings open. Wiener is framed in a rectangle of dirty yellow light. Howard knows how the conversation will begin. *I knew your father*, Wiener will say. And Howard will say, *I know, I'm glad. I'm glad this is going to you* . . . And then? After that? He wishes he had Charlie's skills. But Charlie is dead. He has died of cocaine and freedom.

The ceiling lights flicker on, die, flicker, come on. Harsh light, he would have preferred darkness. Wiener comes slowly up the central aisle of silent machines. He is dignified, like an ageing bear.

— Mr Wiener?

Correct tone, not too careful or over-respectful or too cocky or uncertain.

— I'm Howard Levy.

Wiener looks around. The sight saddens him too.

— I knew your father.

This is going to sound silly to his own ears, he has said it too many times in practice, he has to push on, pretend it was, is, unrehearsed.

— I know. I — I, I'm glad. I'm glad this is going to you . . .

They walk together. They touch machines, they wave unmagic hands around.

— A shame, says Wiener, a waste. This was once so flourishing.

Now. Now he must learn to become like Charlie, seize this moment, if only this moment. With Leslie he was unprepared. This is his chance to pay full tribute to the ancient enemy. The big adversarial power is abruptly gone and Howard feels naked without it and not at all sure what he has been left with. Mark has disappeared, reputedly wealthy. There have been sightings of Simon, too many to believe in any of them, in Hampstead and Bethnal Green and Whitstable and Brighton. The police discovered a forklift truck abandoned in the most overgrown part of the Polish cemetery. Vivien has spoken without result to every hospital and police station in London and the Home Counties.

— My consolations, says sympathetic Harry Wiener, Dinah. Charlie. I'm sorry.

— Thank you.

— Dinah, yes, of course, but Charlie. I knew Charlie when polythene was five pee a pound. Jackie too you know. He died last year. I spend most of my leisure time attending funerals. This is what I tell myself now, everyone goes the same route so how can it be so bad?

Together they contemplate mortality.

— And what, asks courteous Harry Wiener, do you intend to do now?

391

— I don't know. Perhaps travel. I've always wanted to go to America. But also, I wanted to show you these.

Museum Corner. No longer rusting. But primly separated again, back into their feminine machine selves.

— My father's last machines. I had hoped, a favour. I will pay for them of course, if you . . . I would like to keep them, in memory.

— Of course. No payment necessary.

Wiener nods. They shake hands. No Charlie duplicity necessary. The deal is done.

Howard walks the man who knew his father to his Jaguar. He waits in the yard, finds himself a pile of bricks to sit on. Soon the men to take away the machines will be here, to carry them off to his house. And after that . . .? Is there a market for little plastic sex models of red-haired girls? There must be more to the system than that. The dolls are just the teaser. Change the mould and anything is possible. (Even progress?)

Howard waits. He doesn't even know if he has a house to go to. The machines shall travel ahead of him, his very cumbersome calling card. He removes the photograph from his pocket, props it up between half a brick and a splintered plank of wood from a pallet. He cadges a box of matches from a passing workman. Howard hardly winces or smiles as he sets fire to the picture. The first three matches flare up and die and then, the fourth, it catches, just a corner first of all, where the workshop clock shows five to five, then a sudden rich swirl of flame, a puff of blue smoke and it burns.

392